WITHDRAWN FROM
MACALESTER COLLEGE
LIBRARY

Place, Modernity, and the Consumer's World

Place, Modernity, and the Consumer's World

A Relational Framework for Geographical Analysis

ROBERT DAVID SACK

The Johns Hopkins University Press

Baltimore and London

This book was brought to publication with the generous assistance
of the Karl and Edith Pribram Fund.

Published in cooperation with
the Center for American Places, Harrisonburg, Virginia

© 1992 The Johns Hopkins University Press
All rights reserved
Printed in the United States of America on acid-free paper

The Johns Hopkins University Press
701 West 40th Street
Baltimore, Maryland 21211-2190
The Johns Hopkins Press Ltd., London

Library of Congress Cataloging-in-Publication Data

Sack, Robert David.
Place, modernity, and the consumer's world: a relational
framework for geographical analysis / Robert David Sack.
p. cm.
"Published in cooperation with the Center for American Places,
Harrisonburg, Virginia"—T.p. verso.
Includes bibliographical references and index.
ISBN 0-8018-4336-7 (alk. paper)
1. Consumption (Economics) 2. Advertising. 3. Space in economics.
4. Geographical perception. 5. Human geography.
I. Title.
HB801.S23 1992 658.8'35—dc20 92-4391

For Jessica and Joshua

Contents

· II ·
The Consumer's World

· III ·
Geography and Morality

Illustrations appear following p. 96

Preface and Acknowledgments

My published works and teachings have focused on the geographical question of how we make the world into a home. My approach has been to examine the various means by which we conceive of and construct space and place and how these conceptions and constructions affect our lives. The approach ranges from theory and philosophy to description of day-to-day action. One of the great joys of geography—the study of the earth as the home of human beings—is that it offers such a wide and challenging intellectual range.

Three interrelated questions have drawn most of my attention. The first is the means by which physical space acts as a causal agent in human behavior. This led to my work on the relational concept of space. The second is the way in which we conceive of space and place and how their appearance and function change according to our perspectives or modes of thought. This led to my work on conceptions of space. The third is the way in which space and place are constructed and how they form elements of social power. This was developed in my theory of territoriality. Each of these questions interweave space and place with the forces of meaning, nature, social relations, and agency, and with various methodological perspectives. These are the ingredients for making the world into a home, and these inquiries have laid the foundation of a relational framework for geographical analysis.

This book is another step in the construction of that framework. It considers two problems simultaneously. One is the geographical analysis of the consumer's world. Consumption is a virtually universal and accessible means by which we daily create places and construct worlds. Yet the consumer's world is peculiar in the contradictions it contains and in the way in which it severs itself from all other worlds. Consumption creates places that are disorienting and out of place, and yet consumption as a major place-forming activity can be understood most clearly from a geographical perspective.

The other problem is the development of the relational framework for geographical analysis. At a time when modern intellectual debates seem to be wheeling out of control and there appears to be no common ground—when the intellectual world is as disorienting as the consumer's world—it is important to consider the common elements that make the earth into a home. The relational framework sheds light on the nature of the consumer's world and draws attention to the way space and place integrate the multiple perspectives and forces that form the elements of our world. The framework is relational because it is flexible enough to allow the investigator to stand apart and consider a range of forces and perspectives without being committed to any one of them, and at the same time it draws attention to how the world would be if commitments were made to one or another position. The framework is only partially constructed; future projects will extend and direct it toward other types of place and to geographical analysis itself.

The structure of the relational framework bears the impression of my own general philosophical orientation, which is a combination of realism (with a healthy dose of idealism and skepticism) and analytic philosophy. More immediate influences have been the works of Anthony Giddens, David Harvey, Thomas Nagel, J. Nicholas Entrikin, and Yi-Fu Tuan. Giddens's combination of sociology and geography and his thesis of structuration are extremely important to my work. But his "decentering" of the agent and his difficulty in specifying how agents can really have choice leave unaddressed the problem of freedom of choice. Also, his conception of structuration is sociological, and his structures are social structures. Geographers are equally interested in forces from the realms of meaning and nature.

Harvey's analysis of postmodernism is exemplary and has influenced my own, and so too has his analysis of the role of space in Marxist theory. But I am not a Marxist and am reluctant to have a geographical framework that places primary importance on the forces and relations of production serve as the primary approach to geographical questions. These particular components of social relations can come to the fore within the relational framework. But the framework also accommodates the power of other forms of social relations, as well as the complex forces of nature, meaning, and agency. Indeed, the relational framework allows for the possibility that one or another of these forces dominates at particular times, places, and scales.

Nagel's sweeping philosophical analyses have affected my understanding of the role of geography in philosophy. His "view from nowhere," which I have employed throughout the book, demonstrates how the development of awareness is essentially a movement from somewhere to nowhere and how these geographical concepts lie at the core of many philosophical questions. One area that requires elaboration are the forms of awareness and their degrees of abstraction. These are matters that I have attempted to address in this book by considering the axes we move along from somewhere to nowhere.

Entrikin's work on the "betweenness" of place is a compelling analysis of both the capaciousness and elusiveness of place in modern intellectual history. My relational framework takes this betweenness to mean the capacity of place—due to its circumscription of area and to its creation of an inside and outside—both to encompass the forces of meaning, nature, social relations, and agency and to focus our attention on perspectives ranging from somewhere to nowhere. Betweenness is thus the central component of the framework.

Tuan's humanistic geography has had a powerful influence on my own research. His analyses of the interrelationships between space and place, of spatial segmentation and the development of self, and of morality and geography inform each chapter of this book. Tuan reminds us that place is a moral concept, and I share with him the hope that places can become better in a moral sense, though places of consumption can thwart these efforts. Whereas Tuan begins his analysis from the perspective of experience and works outward to concepts and theories, I start in the other direction. Embracing these differences is one of the functions of the relational framework.

The writings of these scholars and many others have helped shape this work. But I have also benefited from numerous long-distance telephone conversations with Entrikin and from face-to-face discussions with Tuan since he joined the Geography Department at Madison. I am also indebted to my other colleagues for providing a most congenial atmosphere for research and teaching. I owe a special debt to Karen Sack, Priscilla Haugen, Paul Adams, Tim Cresswell, David Delaney, John Agnew, and (again) to Nick Entrikin and Yi-Fu Tuan for reading the manuscript and providing numerous constructive criticisms, and to graduate students from my geographical thought classes for doing the the same.

I wish to thank George Thompson for his encouragement and Diane Hammond for her skillful editing. I am indebted to Dave DiBiase for his help in designing the loom illustration, and to Dan Maher for his help in designing the illustration for the relational framework. I also wish to acknowledge the financial support of grants from the Wisconsin Alumni Research Foundation and the Vilas Trust. Finally, I wish to acknowledge that much of this material is based upon work supported by National Science Foundation Grant SES-8604552, though any opinions, findings, and conclusions are those of the author and do not necessarily reflect the views of the National Science Foundation.

Previously published materials of mine have been reworked and included in sections of this book, and I thank the following organizations for their permission to use the material. Revisions of portions of "The Consumer's World: Place as Context," *Annals of the Association of American Geographers* 78 (1988): 642–64, appear in chapters 1 and 6; reworked sections of chapters 2, 5, and 6 of *Human Territoriality: Its Theory and History* (Cambridge: Cambridge University Press, 1986) appear in chapter 2; revised portions of chapter 5 of *Conceptions of Space in Social Thought: A Geographic Perspective* (Minneapolis: University of Minnesota Press, 1980) appear in chapter 2; reworked portions of "Social Theory and Environmental Research," in B. L. Turner, ed., *The Earth as Transformed by Human Action* (New Haven: Yale University Press, 1990), appear in chapter 4; and several of the ideas for chapter 8 were introduced in "Strangers and Places without Context," *Annals of the Association of American Geographers* 80 (1990): 133–35.

Reproduced from the *New York Journal*, 9 July 1772: figure 3. Reproduced from *Furnishings for the Middle Class: Heal's Catalogues, 1853–1934*. New York: St. Martin's Press, 1972: figures 4 and 5. Reproduced with permission of Hiram Walker and Sons, Inc., Detroit: figures 10 and 11. Reproduced with permission of Rodier Corporation, New York: figure 12. Reproduced with permission of Hill, Holliday, Connors, Cosmopolus, Inc., Advertising, Marina del Rey, California, and Greta Carlstrom, Marblehead, Massachusetts, photographer: figure 17. Reproduced with permission of U.S. Virgin Islands Division of Tourism, New York: figure 21. Reproduced with permission of Levenson, Levenson, and Hill Advertising Agency, Dallas/Fort Worth International Airport, Texas: figure 24. Reproduced with permission of Jill Posener, photographer: figure 30. Reproduced with permission of West Edmonton Mall, Edmonton, Alberta, Canada. Photograph by Gordon Cook: figure 33. Photographed by Paul Adams: figure 34.

Place, Modernity, and the Consumer's World

· 1 ·

Introduction: Places of Consumption and the Relational Framework

The premise of this book is geographical: that space and place are fundamental means through which we make sense of the world and through which we act. I use this premise to show that places created by and for mass consumption are fundamental to our making sense of the modern world (and of one of its facets, the postmodern) and to our power as agents in that world. To support these claims, I examine the consumer's landscape and its role in contemporary life. I also extend and elaborate the geographical premise concerning the significance of space and place. Such an elaboration provides the outline for the discussion of a relational geographical framework (illustrated in figure 1), which explores the power of place in the modern consumer's world and the overall characteristics of geographical space and place and their relations to experience and action. The book moves back and forth, in both directions, at one moment addressing the significance of the consumer's landscape to modernity, at another moment extending the premise to outline a framework for the role of space and place in human action and experience.

Consumption and modernity, on the one hand, and the relational geographical framework, on the other, are mutually reinforcing. Consumption is basic to living in the modern world. Even though we differ from one another in many respects, it is a fact of modern life that most of us are consumers and that we share the experience of being in places of consumption. Places of consumption are central to modernity; both need a frame of analysis broader than they are—a relational geographical framework that can be applied to the current character of space and place and also to their role in other historical contexts. The completion of this framework and its applications to a wider range of geographical problems must await a subsequent work. Here it is sufficiently developed to serve as a map, providing a set of coordinates and projections that can help us locate and analyze a variety of geographical places and practices within an intellectual terrain of perspectives and forces (including meaning, nature, social relations, and agency). The framework reveals how these forces, perspectives, and places are mutually constitutive. This dynamic and relational quality means that particular geographical practices not only fit within this framework but rework it. These relational qualities are illustrated by the geographical model of consumption (figure 2), which describes the process by which consumption creates places that simultaneously weave together and alter forces and perspectives.

Most people probably are better acquainted with mass consumption than with geographical analysis, so I begin by examining places of consumption. But first, some terms must be clarified.

Space and *place* are extremely complex and interrelated concepts. Space can be a set of places, and place can be a location in space. A major point of this book is that the meanings of space and place depend on the particular perspectives from which they are viewed. Another set of distinctions is that between *place* and *landscape* and between *place* and *context.*

Landscape is the visible quality of place. A shopping mall or a resort is not only a place in which things are consumed but also a place whose landscape is arranged to encourage consumption; and indeed, the appearance of that place—its landscape—is often the element that is consumed.

Sometimes places in the consumer's world are ephemeral and loosely organized. This is especially the case in advertisements and in some store displays that present the commodity in a vague and fleeting background or place. I use the term *context* to indicate this ephemeral quality of place, or I join the terms together in "place as context." Other nuances and meanings of space and place are introduced as we go along.

Place, Consumption, and Modernity

Places of consumption have rapidly spread across the landscape in the last hundred years. They constitute much of the modern home and its furnishings, planned neighborhoods and housing developments, shopping strips along highways, cityscapes, shopping malls, recreational areas and resorts, recreational theme parks, and natural settings, and vast tracts of countries that are mass consumed as tourist attractions. Most of our nonworking lives are lived in such places—created by and for the consumption of mass-produced products—and our exposure to them is reinforced by the mass communications media. Place and consumption are connected—not simply for the obvious reason that we must consume things in place but in the more important sense that consumption is a place-creating and place-altering act.

Mass consumption, in fact, is among the most important means by which we become powerful geographical agents in our day-to-day lives. (The term *geographical,* in the field of geography, does not refer to the material or natural environment only, but to the social and the intellectual environment as well.) As consumers, we are capable of altering these environments simply by being links in the production-consumption chain: each act of consumption draws on products whose very existence is based on the extraction of raw materials, their assembly in purposely built factories, and their distribution by way of complex transportation systems. Consumption is part of all these processes; each act of consumption affects these processes and their landscapes in a real way and, in addition, creates new places with new effects. Yet these places of consumption, in fact, attempt to sever their connections to these other processes and places by presenting themselves as a world apart—a consumer's world, a showcase of goods and services, tours, and vistas. The consumer's world and its places or landscapes of consumption emphasize only those activities directly linked to the purchase and use of mass-produced commodities or to the mass consumption of places and vistas. By severing our connections to the rest of the world, places of consumption encourage us to think of ourselves not as links in a chain but, rather, as the center of the world. Each time we bring a product home, our home is transformed physically—something is there that was not there before—and transformed in terms of values and meaning. Bringing home a Rolls Royce means something different from bringing home a Honda Civic. The same transformative effects hold for commodities that furnish workplaces and places for recreation. Commodities also affect the structure and quality of the places in which they are sold. Department stores, shopping malls, and urban shopping centers are designed to create appealing contexts to enhance consumption: they become stage settings and even spectacles. These concentrated forms of the consumer's landscape involve discrete commodities. But virtually any place can become commoditized—and consumed en masse: not only can a place or an area become a parcel of land that is bought and sold over again in a market economy, but the experience of being in that place can be "sold" or "rented" to consumers (and this does not necessarily require that the place be owned privately). This kind of commoditization of place occurs with theme parks, tourist sites, and even countries and continents.

The place-creating and place-altering power of consumption extends beyond places of consumption to include commodity production, commodity distribution, and the pollution caused by both of these processes.

But by severing its connection to these consequences and presenting itself as a world apart, the consumer's world magnifies and distorts this particular power of the consumer to create places.

These place-building qualities of consumption resonate throughout our culture with the primary assistance of advertising. One of advertising's primary messages is that commodities are place creating: it tells us that a commodity can help us create a context, a place, or a world, with us at the center. If there is the least doubt that advertising is a geographical map showing the contexts or worlds that commodities are meant to create, consider the gallery of illustrations in this book. Even though we may be skeptical about advertisements, they provide an idealization of how commodities are supposed to work and thus are a basic component of meaning in the consumer's world.

Mass consumption is, then, a place-creating activity. But of equal importance is the fact that essential qualities of commodities cannot be understood unless they are seen as place creating. When feelings about the world become part of a place, they become concrete and natural: they become real qualities.[1] And what are the attributes of these places of consumption? The answer is, they contain the same dynamic and contradictory attributes that characterize modern life. These places have been described as cornucopias, wonderlands, and paradises. But they have also been called disorienting, weightless, and inauthentic spectacles that destroy real contexts, creating illusion and diversion. They free us from necessity, yet they cut us off from responsibility. They are exhilarating and disorienting. They make our contradictory feelings about self and world, stability and change, more concrete and visible—in places that, paradoxically, are themselves constantly changing and disposable. These places, along with everything else modern, evoke Yeats's image of a world revolving around a center that "cannot hold" and Marx's image of a world "pregnant with its contrary . . . [so that] all that is solid melts into air."[2]

Place helps make feelings real.[3] Experiences and ideas have immediacy, but they are impermanent without place and its artifacts to anchor them. As Yi-Fu Tuan argues, "transient feelings and thoughts gain permanence and objectivity through things,"[4] and these landscapes or places become repositories of meaning. But when places of consumption themselves evoke the feeling that they are unreal, magical, impermanent, and inauthentic, then the very grounds for experiencing reality are shaken.

This connection between landscapes of consumption and the modern paradox is evident historically. Premodern agrarian societies forged close connections between meanings and actions, on the one hand, and places, on the other. Things and events were rooted in place. This intimate connection was experienced personally and shared publicly. The most frequent exception to this fusion of thing and place was trade and commerce: the uprooting of things and their movement from one place to another, which constitutes trade, was associated with the merchants themselves, who were often thought of as rootless. Still, their activities were usually contained within urban centers and, even then, in special areas or enclaves; they did not overwhelm the rural landscape until the rise of modernity.[5]

Although trade and commerce became truly global in the fifteenth and sixteenth centuries, geographical transformations were gradual and unevenly dispersed over the landscape until the eighteenth century, when manufacturing and production spread. From the eighteenth century until the twentieth century, places of production (mines, mills, factories, and the industrial city) were powerful geographical manifestations of modernity. In this landscape, the workplace was spatially distinct from the home and presented a powerful image of modernity, while the home place remained a powerful image of the traditional.

In the twentieth century, consumption became the locus of modernity, spreading beyond the home and the local and geographically discrete retail establishments to create vast landscapes of consumption—revitalized downtowns, shopping malls, and recreational and tourist attractions. A consequence has been to make the effects of modernity visible in all facets of life: we become more conscious of modernity when the places we occupy in our daily lives are the very loci of the contradictions of modernity.

Modernity

Modernity contains many components, and it is hard to say which one is basic. Certainly its economic component is extremely important, but other social relations, science and technology, and values, attitudes, and beliefs are also powerful forces.[6] An important theoretical issue of modernity in particular, and of human behavior in general, is deciding the relative importance of these components. Strong claims can be made that attitudes, values, and beliefs (what I call *meaning*) are as important to modernity as the economic component. Indeed, one of the distinctive characteristics of modernity is its self-consciousness or reflexivity, its awareness that specific meanings and feelings are associated with modernity. The term *modernity* combines social processes (modernization) and intellectual processes (modernism) and embraces our self-consciousness about the modern world. Modernity is an extremely reflexive process, whose latest and most self-conscious expression is found in the condition of postmodernity.[7]

According to Marshall Berman, the history of modernity can be divided into three phases, which correspond to the development of the modernization process itself.[8] In the first phase, which spans roughly the period between the beginning of the sixteenth century to the end of the eighteenth (the period covering the rapid globalization of trade and commerce), people were barely conscious of the nature of the change that began to envelop them. In the second phase, from the late eighteenth century to the beginning of the twentieth (a period spanning the industrial revolution), the modern public emerged and became aware of living in a revolutionary and contradictory age and yet could remember what it was like to live in a quiet, premodern world. "From this inner dichotomy, this sense of living in two worlds simultaneously, the ideas of modernization and modernism emerge and unfold."[9] In the last phase (the twentieth century), these contradictions spread worldwide. They are expressed in

> our desire to be rooted in a stable and coherent personal and social past, and our insatiable desire for growth—not merely for economic growth but for growth in experience, in pleasure, in knowledge, in sensibility—growth that destroys both the physical and social landscapes of our past, and our emotional links with those lost worlds; our desperate allegiances to ethnic, national, class, and sexual groups which we hope will give us a firm "identity," and the internationalization of everyday life—of our clothes and household goods, our books and music, our ideas and fantasies—that spreads all our identities all over the map; our desire for clear and solid values to live by, and our desire to embrace the limitless possibilities of modern life and experience that obliterate all values.[10]

The contradictory qualities of modernity that make "all that is solid melt into air" now threaten to split the intellectual world into fragments that oversimplify conditions and forge "rigid polarities and flat totalizations."[11] Three of these oversimplified and fragmentary positions are extremely important geographically: modernity as optimistic and global, modernity as nostalgic and local, and the reaction of postmodernity.

Modernity as Optimistic and Global

The first position is optimistic about modernity and equates modernity with progress. This position has confidence in modern science and technology and sees the possibility of a rational world order that can sustain a high standard of living for the entire world, a world virtually free of necessity, perhaps a consumer's paradise. This optimistic and global position disdains the past: this older world contained many local economies and communities, which encouraged narrow interests and parochial moral systems and chained people to place. Modernity ushered in a global economy and a global culture, which promises world citizenship and universal moral principles. Such a world must be planned and cared for, and its landscapes must contain scientifically efficient land use and transportation systems, based on a public view of space—one that is abstract and geometric.

This view has us regard distance as a cost to interaction or transportation and regard place as a location containing bundles of attributes that could also occur elsewhere. It provides a unifying framework, but it has little capacity to record our feelings about place and our experience of being in them. It does not dwell on the unique character of each place. This abstract view of space suspects any interest in the specificity of place and the local community, which it sees as leading to a conservative and parochial outlook and an impediment to global planning. This optimistic, global position emphasizes the exhilarating effects of space, of movement, and of the weightlessness of modernity.

Modernity as Nostalgic and Local

The second position is pessimistic about the global thrust of modernity and believes instead in the virtues of the local community. It, too, begins with the assumption that modernity has created a shift away from local community and its traditional sense of place and has weakened important institutions. Geographical homogeneity has increased, reducing the number of social and environmental experiments and weakening human resistance to natural and human-made hazards. The local community has been undermined, which, it argues, is the natural (and perhaps the only) basis for moral order. Taking the place of the local community are many territorial communities of convenience, which lack true purpose and cohesion. The results are social systems containing isolated individuals, who do not know their domains of responsibility and the consequences of their actions and who therefore direct their efforts toward their own gratification. A remedy, according to this position, would be a return to local communities, in which we can become rooted.

Postmodernity

The two positions—one future oriented and global, the other nostalgic and local—oversimplify and emphasize only a portion of the tensions of modernity but are, nevertheless, confident in their suppositions about how the world works and what can be made of it. Both then are products of the Enlightenment project, as it has been called, which exudes a confidence in rationality and in the belief that there is a truth to be known and a position from which to see it. This is why each position has engendered political movements and utopian plans.

A third reaction to the contradictions of modernity is postmodernism.[12] This intellectual stance (associated with the more general and putative condition of postmodernity) attempts to swim in the turbulent currents of contemporary culture without rising above them to determine their structure and their connection to the rest of the world. Postmodernists embrace a radical relativity and despair of the possibility of finding truth. This skepticism has been fueled by the contradictions between the first two positions and by a too literal reading of philosophies and histories of science that point to the impossibility of objective and certain foundations of knowledge—a reading that, interestingly enough, has not impeded the development of the natural sciences, since they see scientific enterprise as a constant revision of models of a real world that itself cannot be known except through our senses and thoughts.[13] Still, doubts about the absolute foundations of scientific knowledge have been used by many in the social sciences and humanities as justification for the relativity of virtually all perspectives, and this has added to postmodernism's radical skepticism.

Skepticism is an ancient philosophical position and arises from philosophical paradoxes. Given the contradictions of modernity, it is not surprising that skepticism is attractive. Indeed, the contradictions between idealism and realism have led to a healthy and creative skepticism.[14] But postmodernity has also led to a far more radical form of skepticism. Faced with a paradoxical world, denying that science can use rationality to find truth, and doubting that we can construct small, cohesive communities that would stabilize the chaos and achieve a "local truth,"[15] advocates of postmodernity are left with a narrow range of skeptical possibilities, from nihilism (which exalts human actions) to permissiveness (which sees no possibility of objective truth). As David Harvey puts it:

> If, as the postmodernists insist, we cannot aspire to any unified representation of the world, or picture it as a totality full of connections and differentiations rather than as perpetually shifting fragments, then how can we possibly aspire to act coherently with respect to the world? The simple postmodernist answer is that since coherent representation and action are either repressive or illusionary (and therefore doomed to be self-dissolving and self-defeating), we should not even try to engage in some global project.[16]

Postmodernism is not just a reaction to intellectual contradictions: it is equally a response to tensions in our everyday experience of being in the modern world. These tensions are heightened in certain kinds of modern places, such as the landscapes of consumption that combine and intensify the strands of modernity. Landscapes of consumption stand in an intermediate and mediating position between places that are future oriented, dynamic, and have a scientific sense of space—the highly geometrical and well-ordered landscapes of transportation and settlement—and those that provide a more traditional, local, and concrete sense of place—which in a world of strangers threaten to become personal, idiosyncratic, and subjective. Landscapes of consumption become microcosms of the same contradictions that postmodernists take to be the conditions of reality. A single place can be dynamic and liberating, an image of a paradise, and also shallow and disorienting, an inauthentic pastiche. Such places, like modernity itself, cut "across all boundaries of geography" and history and provide, paradoxically, "a unity of disunity." They become places of "perpetual disintegration and renewal, of struggle and contradiction, of ambiguity and anguish."[17]

If postmodernism were only the observation that such sensations exist in the modern world, we would take its contributions to be a description and even a playful parody of these sensations. Postmodernism does not ask how such places and sensations have come about; on the contrary, it denies the conditions of a coherent answer and immerses itself in the experience, taking it to be the only reality. From a geographical perspective, postmodernism assumes that the consumer's world is total. To David Harvey, "postmodernism swims, even wallows, in the fragmentary and the chaotic currents of change as if that is all there is."[18]

And so postmodernism finds no firm ground on which to stand. When it attempts to produce ideas or build buildings, they too are disorienting and fragmentary—a collage or pastiche or montage.[19] "Fiction, fragmentation, collage, and eclecticism, all suffused with a sense of ephemerality and chaos, are, perhaps, the themes that dominate in today's practices of [postmodern] architecture and urban design."[20] It is postmodernism as an intellectual orientation, rather than postmodernity as a type of modernity, that I and many others find unacceptable.[21] Indeed, one of the points of this book is that such disorienting and contradictory feelings about the world are part of modernity and that they can and must be analyzed. To do so, one must get a feel of the conditions but also stand above them. One important analysis argues that both modernity and postmodernity are linked to particular forms of social relations: that postmodernity is, in Frederic Jameson's terms, a form of late capitalist cultural production and, in Harvey's terms, a cultural manifestation of flexible accumulation.[22] Since I am focusing on consumption, one might think that I, too, link modernity to specific social and economic forces. But although these certainly are part of the conditions, consumption, commodities, and indeed capitalism itself are grounded in far broader issues. Therefore, a fuller picture must be developed by extending the geographical premise about the role of place in experience and action.

For now, though, it should be clear that places of consumption are a basic property of modernity. They shape everyday life and provide us with the setting to become agents in the modern world. This claim, of course, is embedded in the more general geographical premise that place is a fundamental means by which we make sense of the world and through which we act. But this general claim about places of consumption leads us to ask if there are not other kinds of place that are equally important to modernity. Why not focus on the producer's world, or the world of work, or a geographical entity like the city? The answer, of course, is that we can indeed gain insight by investigating these and other kinds of geographical forms and activities, and several of these will be pursued later. Our landscapes of consumption, after all, are linked to these other worlds, because consumption is part of a complex production-consumption chain. Furthermore, no place exhibits only one kind of process, and landscapes of consumption are no exception. Even though such places are dedicated to encouraging consumption, they are also workplaces for the employees selling the products and creating the environments. Retailing, tourism, and entertainment are self-designated industries.[23] These facets of place, as well as consumption's connection to other facets, are not ignored.

Still, concentrating on consumption is justified because it is in these landscapes that the most geographically diverse and quintessentially modern experiences are to be found. In addition, these places are extremely accessible. World culture is inextricably linked to world consumption and advertising. Virtually everyone in the modern world, no matter how reluctantly and disdainfully, is of necessity a consumer and aware of the consumer's world, even if not inhabiting it fully. Moreover, as consumers of mass-produced goods, we all share the fact that we are (unwitting) agents jointly transforming the world.

For these reasons, I concentrate on places, or landscapes, of consumption and support the claim that they are basic to modern life by demonstrating that their structure contains the tensions of modernity. But what do I mean by *structure, tension,* and *place?* To give these key words (which are also words of common and loose usage) greater precision and comprehensiveness, and to demonstrate the central role of places of consumption to modern life, I turn now in the other, more abstract, direction

and elaborate the general geographical premise—that space and place are basic to experience and action. Without such an elaboration, explanatory sketches of the modern home, shopping mall, and Disney World—in short, the modern world of consumption—would always seem ad hoc and arbitrary.

Place, Experience, and Action

My discussion has already suggested that geographical places help constitute and are the products of many processes and exist in many forms and scales. Places of consumption are a fairly technical type of place, one that does not jump to the general public's mind when the word *geography* is mentioned. More familiar to the general public are those types of places listed in gazetteers: political units, cities, regions of the world, oceans, lakes, continents, mountains, deserts, forests. By virtue of their proper names, each of these places can be thought of as a unique area, region, or place. (The three terms are often used interchangeably, although there is some implication about scale, with region being the largest area.) The various characteristics of a place emerge when it is examined in certain lights. We can, for example, analyze place in terms of the processes it contains and constitutes. Cities would, in this light, break down into retailing or manufacturing sites or residences. The upper Midwest would be thought of as a corn-producing region or as part of the old industrial belt. Processes and their places also include underdeveloped areas and ghettos, stream dynamics and drainage basin morphology, soil and biotic regimes, transportation and highways, work and factories, and consumption and its landscape.

Geographers also examine place in terms of how the people who live there construct, sustain, perceive, and experience it. These perceptions may be quite different from those of outsiders; the places that people identify as important to them, such as their local neighborhoods, may not be recognized as places by outsiders or by geographers. Because we live in a geographically mobile society, divergences between the outsiders' and insiders' views of a place are often great.

Individual views of place have even less chance of being reinforced and shared by others than group views, so these tend to be personal, subjective, and even idiosyncratic. Yet these personal views are the most elementary and accessible way of being in geographical space and place, for it is the way we constantly experience the world and act upon it. Being in the world is being in, and constructing, this personal sense of place, with ourselves at the center.[24] Personal place expands and contracts as our interests and actions wax and wane. And personal place moves as we move through space. In a closely knit and geographically isolated community, the contents of our personal place can be shared by others and can coincide with (or develop into) a place fixed in space and publicly recognized. But in a fragmented and dynamic society, personal place is less likely to be shared and so can appear private, idiosyncratic, and subjective.

Place, then, is complex. In the personal case, place moves as we move. Movement is one of the ways place differs from space.[25] Space is the stable framework that contains places, and the geometrical structure and precise coordinates of space makes it possible for us to notice that a place can move because it shifts its locations within space. Over a long span of time, all of the places I describe above move, all of the activities they contain and their relations with their surrounding places can shift locations: cities, ghettos, tropical rain forests, and even oceans and continents move.

The development of abstract space that can be described geometrically is relatively modern and Western. The system is based on developments in cartography, perspective painting, mathematics, and territoriality.[26] Such a system, with its mathematical coordinates cartographically expressed in lines of longitude and latitude, fits the needs of a mobile and technical society and has become the public, objective conception of space. With its adoption, the basic meaning of *place* became *location in space;* all of the other meanings of *place,* though still differing from one another, now also contain this meaning, as well.

Place and Perspectives

Space and place, then, are varied and inescapable parts of experience. By their very nature, they raise issues that lead to a particularly modern quandary: the debate about the proper perspective for examining things. Are places best seen, for example, from a positivistic social science perspective or from a humanistic perspective?[27] Issues of perspective are extremely important to contemporary debates about knowledge, for it is possible to argue that what is there depends on perspective, and the modern world contains a proliferation of perspectives from which we view things.

Space, place, and perspective are obviously interrelated. A perspective can be literally a view from a particular place, and so multiple places provide multiple perspectives. But the relation goes even deeper. Place is bounded and thus can be seen literally and imaginatively from within and from without. There are degrees of "outsideness." The literal and imaginative inside/outside (with the latter's many degrees) blend together and form a key component of awareness. This happens because literal spatial manipulations (as we shall see) are at the core of imagination. These movements from inside to outside correspond to Thomas Nagel's more general distinction between "somewhere," as personal perspective, and the various degrees of the more public, abstract, and objective "somewhere else," which lead to the most abstract virtual "nowhere."

This inside/outside attribute of place makes it one of the few categories of thought that demands the simultaneous involvement of more than one perspective, and these perspectives can be turned on the place that generated them. That is, each place can be viewed, through the imagination, from within and from without; and views from outside can vary in abstraction from being in place to being virtually nowhere. The geographical imagination has always dealt with the entire range of perspectives, from somewhere to practically nowhere. Modern philosophy uses these perspectives to array its philosophical positions. Nagel, for example, uses these perspectives to organize epistemological issues from science to morality to art, wherein the somewhere is a more personal, subjective, position, and the nowhere is a more abstract and objective one.[28]

Place, and its perspectives from somewhere to virtually nowhere, is the organizing system for the development of the human intellect. The growth of awareness from birth to adulthood has been described by Jean Piaget and others as the capacity to first perceive, and then conceive of, ourselves from different perspectives in space (first from a topological, then from a projective, and finally from a distant and abstract Euclidean perspective).[29] These spatial operations become embedded within all other mental abstractions, which coincides with the common understanding of mature awareness as the ability to see things abstractly and from the viewpoints of others. Space and place, then, are essential to awareness. However, modern places of consumption alter perspectives to such a degree that they create a sense of disorientation.

Place and Forces

Space and place are central to another realm of issues: the primary forces that affect our lives. Whereas perspectives are about how to see things, forces create the things we can see. The forces that affect us encompass factors that contribute to our actions, including ourselves as causal agents. A basic question is the degree to which we, in ourselves, can be causal agents rather than simply agents of other causes or forces—which can be either outside of ourselves, as in natural or social causes, or within ourselves, as instincts or as psychological forces (including attitudes, values, and beliefs). It is one of Anthony Giddens's great insights to have seen that forces (or their manifestation in structures) are instantiated and perpetuated through behavior and that these forces affect us by constraining our behavior or by enabling us to act.[30]

While of enormous importance, this duality of forces or structures does not provide a means of distinguishing agents as forces from agents as vehicles of other forces. It "de-centers" the agent and leaves no room for choices based on free will or the ability to act otherwise—"reflexive monitoring" notwithstanding.[31] (This limitation is addressed in chapters 2 and 3.) Rather, structuration can be understood as a series of forces or structures that are constituted and also altered through behavior and that enable agents to act as vehicles of other forces or structures. (This does not address the problem of free choice; see chapter 3). This meaning of structuration is assumed throughout most of this analysis, so that when I discuss forces, I assume that they are instantiated and even constituted by and through behavior. But I also consider the more general and philosophical question of free will or free agency, because the belief in agents who are themselves causes or forces, and not simply vehicles of other causes or forces, is essential for moral responsibility.

Given this understanding, the modern period conceives of these forces as stemming from four somewhat separate and distinct realms. One of these is the *realm of nature.* The term *nature* connotes many things, but in the contemporary world we often turn to the natural sciences to understand what nature is about. In this respect, the realm of nature becomes the elements and forces described by the natural sciences and this is how the term is used here. There is no doubt that we are affected by these forces; theories from physics, chemistry, and biology tell us how we move, how we metabolize, and how we reproduce. But we are also influenced by forces that we ourselves produce or, more precisely, that appear to be of our own making. These have been referred to by the general term *culture*, so that, at the most abstract level, we are products of nature and culture. (See chapter 4 for definitions of *nature* and *culture.*)

Culture consists of two primary and equally important loci or realms of forces. One combines social, economic, and political forces and is called the *realm of social relations*. The other combines the ideas, values, and beliefs that give meaning to the world, and is called the *realm of meaning*. A fourth realm of force—the *realm of agency*—goes beyond the structuration de-centered position regarding agents; it claims that forces from the other realms are not determinate, precisely because we have power over ourselves. This power is called free will, or free agency.

It is extremely difficult to isolate and is not normally considered apart from the other three.

Specific theories focus on how these forces from one or another realm exert power over us. A basic assumption of the natural sciences is that humans are part of the natural world and thus governed by all of its forces. The realm of social relations investigates how factors such as class, mode of production, and bureaucracy affect our behavior. Theories of meaning that involve personality and symbolic structure describe how our ideas, values, and beliefs have particular forms and influence our actions. At first glance, this division roughly corresponds to the academic one of natural science, social science, and the humanities. Yet there are important distinctions between the three realms and these academic terms.

Forces from different realms are difficult to interrelate in the abstract, even though most phenomena of everyday life are composites of these forces. Most theories emphasize forces from only one of two realms. Natural scientists spend most of their time researching natural processes, not their connection to social relations and meaning. Social scientists usually consider such social relations as bureaucratic or class structure or monetary policy, and pay less attention to people's attitudes or to the natural world. And intellectual historians do not emphasize social relations or the natural world. Needless to say, most of the phenomena of everyday life are composites of these realms. It is intellectual life that separates them. On those occasions when theories from one realm consider the others, they emphasize how their realm dominates the others. Theories of environmentalism and sociobiology claim that our social organizations and our meanings (ideas, values, and beliefs) are driven by natural forces; theories of meaning emphasize how our ideas, values, and beliefs structure society and "construct" nature; and theories of social relations sometimes claim that a society's ideas, and even its conceptions of nature, are forged by economics and social organization. Since all of these theories can be plausible in particular circumstances, but are generally incomplete and only partially specified, the intellectual community is at something of a deadlock.

The difficulties in integrating these forces are compounded by the fact that we must use specific perspectives to examine them and that these perspectives, in turn, affect what we take these forces to be. How to integrate forces and perspectives is the major theoretical problem of modern social theory. It is at this juncture of conflicting forces and perspectives that space and place enter as essential categories. Being inside and outside of place invokes more than one perspective; hence, place can provide a basis for examining conflicting points of view. Place incorporates

forces from all of the realms, though not always in the same proportions. The capacity of place to integrate forces may not be as obvious as the capacity of place to integrate perspectives, since our society has so fragmented the forces that we tend to think of place as a product of only one or another of them. Thus we speak of wilderness and tropical rain forests as natural places, ghettos and slums as social places, and personal and holy places as products of meaning. This is not incorrect, if we have identified the principle force that constructed such places. But the character of a place usually involves all of the forces: place is one of the few categories where all the forces meet. Consider so-called natural places, such as tropical rain forests or national parks. Although they do undoubtedly contain natural forces, they are partially the products of political jurisdictions and other forms of social control that protect them from the ravages of civilization, and they are also the products of the meanings we impart to them.

In a general sense, places are partly intellectual constructs. (This claim is not a commitment to idealism; see chapters 2 and 3.) Places are seen and understood because our minds delimit and categorize the phenomena of the world. Wilderness and tropical rain forests involve meaning in the direct sense that they embody particular values—we prize these places and wish to preserve them. Places that are normally thought of as social are also confluences of forces from other realms. Ghettos and slums contain not only social relations but also physical structures, climate, and physical distances, as well as meaning: ghettos and slums are, to many, undesirable places. Similarly, even our own personal places are confluences of forces: we are surrounded by such natural phenomena as air and gravity; we are certain physical distances from other things; our rooms, houses, or libraries are socially supported; and our places have meaning to us as loci of attention or "fields of care."[32] Finally, place requires human agency, whether as a vehicle of mental, social, or physical forces, or as a source of power in itself.

Place, then, is a basic and accessible category of everyday life that incorporates forces and perspectives. It is involved in any action we undertake, and our actions create and alter it. This is the general geographical premise. Still, it does not tell us how actual places such as landscapes of consumption are constructed and how they function.

The Relational Framework

A formal picture of the relations among these forces, perspectives, and places can provide a conceptual map of the modern intellectual terrain (see figure 1). It allows for a temporary suspension of commitment to current theories and viewpoints, so that a common ground can be explored, and yet it does not join place or space to only one perspective or force or succumb to the postmodern urge to abandon all frameworks. Rather, it provides a multifaceted and relational view, allowing interplay among forces and perspectives. Portions of these relations have been elaborated in the geographical literature. Here, this relational framework is developed enough to shed light on the general power of place and on its particular connections to modern debates.

The categories of the framework are not absolutely distinct but are permeable and overlapping, and their conceptual separation itself has an important history. The overlapping realms at the base of the diagram contain the forces of meaning, nature, and social relations. The cone rising from the center contains the axes of abstraction, along which are located well-developed positions or perspectives (indicated by the lenses). The relations among these ontological and epistemological categories are discussed as I elaborate the framework, but two connections are important to mention at the outset. First, agency is not shown as a separate realm, because the sense that the agent has real choices or can "act otherwise" diminishes as the perspective moves along the axes from somewhere to nowhere. Rather, agency pervades the realms as a constitutive element because of Giddens's "duality of structure."

Second, the connection between perspective and the realm of meaning is pivotal and intimate. Meaning is a mental structure that guides our actions and affects how we see the world. In this sense, any set of meanings can serve as both a force and a perspective. But only a few meanings—such as those of science, art, and ethics—are comprehensive enough to produce self-consciousness. These meanings (which are aware, reflexive, and comprehensive) I call *perspectives*. In other words, these meanings provide perspectives on the other forces as well as on perspectives. Conversely, a perspective can be both a view of the world and a force that helps shape it, because a perspective is a set of meanings that guides our actions. Science, for example, has been discussed both as a perspective for viewing the world and as a force that shapes our attitudes, values, and actions. The same can be said for religion. The consumer's world comprises a set of forces, but only in a limited sense does it constitute a set of perspectives. It is precisely because the consumer's world view is so limited and unreflexive that it does not constitute a perspective in the sense I use the term here.

Figure 1 illustrates several of these relations. General perspectives—including aesthetics, morality, and the discursive, scientific mode—are each represented by an axis extending up from the center (the somewhere) to the more abstract position of virtually nowhere.[33] The fact that only four axes appear on the diagram signifies nothing more than that four is a sufficient number to convey the range of possible perspectives. By the same token, no particular assignment of perspectives to axes is intended, nor is there any particular significance to the inclinations of the axes or the number and locations of the lenses.

The cone is intended to suggest two things: (1) somewhere and nowhere are limiting cases and are never in themselves completely attainable, and (2) there are many paths from somewhere to nowhere, and yet they all are interrelated. The lenses on these axes indicate methodological positions within these perspectives. For morality, emotive ethics would specify a moral view (or lens) close to somewhere. At the other extreme, a utilitarian calculus of the greatest good for the greatest number would specify a view (or lens) from nowhere. For the discursive, scientific axis, a humanistic social science perspective (which tends to be personal and subjective) would be a lens near somewhere. At the other extreme, a positivistic conception would push science toward a view from nowhere (which has been described as a public, objective perspective).

Some things can be seen more clearly from one axis than from another or from a particular lens along that axis. In general, we both gain and lose as we move from somewhere to nowhere. Moreover, the framework allows us to imagine what would happen if one or another perspective were used to examine space and place. For example, along the discursive, scientific axis, an abstract lens close to nowhere would draw attention to place as location in space, whereas a less abstract lens along the same axis from close to somewhere would draw attention to the personal sense of place.[34]

The framework is relational because it suggests how the perspectives allow us to see the different properties of forces and because it draws our attention to how these perspectives are based on forces. Figure 1 is intended as a sketch, a map of the general relations among forces and perspectives. This map can be developed further by detailing various axes, their lenses, and their orientation to forces. Filling in these details may rearrange the map, by imposing a particular perspective or emphasizing a particular force. Thus extension of the framework can result in numerous subframeworks. This is in fact the case with the model of consumption illustrated by the loom (figure 2). Here, the everyday act of consumption addresses the theoretically difficult problems of the framework by drawing together elements of forces and perspectives to construct place, while simultaneously altering the character of these forces and perspectives.

The center of the map in figure 1—defined by the overlapping realms of forces and by somewhere (the point of departure for the axes)—

contains geographical space and place. But these are also present in each of the realms separately. Space is a component of nature: meanings, social relations, and actions occur in space and are affected by it. Without addressing the perhaps intractable issue of the degree to which each of these realms contains the same sense of space as is found in the center, I note that space is a part of each but that claims have been made that space is essential and even constitutive of the realms. That is, it is an ontological category. For instance, it enters as a necessary component of meaning in Immanuel Kant's synthetic a priori and in P. F. Strawson's claim of the indispensability of space to the process of individuation.[35]

Space is also an ontological category in nature, whether in the form of a Euclidean space (for a Newtonian world) or a non-Euclidean one (for a relativistic world). Space is ontologically fundamental to agency, in that it is part of action, and constitutes the experience of being in the world.

The least clear role for space as an ontological category is in social relations. Edward Soja's use of "spatiality" only reiterates the desire to have it become a basic category, and David Harvey's insertion of the spatial into Marxist analysis still does not make space as basic a force in Marxism as class relations.[36] Still, space enters social relations because these must rely on the natural world and its rules and thus must use physical space, although relationally. Social relations must also draw on the realm of meaning and thus must conceive of space as a category. But this appropriation of space makes social relations, from the ontological view, less basic than other realms. This is not a satisfactory position for many social scientists. Perhaps the most fundamental concept of space in the realm of social relations comes from the theory of territoriality, which describes how and why territories help bring influence or power into being.[37] Territoriality is assumed, for example, in much of the discussion of spatiality, and in the formulations of David Harvey and Michael Mann.[38] Though territoriality draws also on the realm of meaning, it nevertheless is intimately involved with a general sense of social relations, insofar as power is a property of such relations.

This then is the extension of the geographical premise to the point where it leads to a framework that can be used to discuss the central roles of geographical space and place. My particular interest is the manner in which the geographical analysis of places of consumption fits within this framework.

A Geographical Model of Consumption

By simply being in space and creating a place, we become involved in the relational framework. Places of consumption are involved in a particularly important way. The geographical model of consumption (figure 2) shows how consumption weaves together and alters elements of the relational framework. In the modern period, the perspective from somewhere coalesces with a personal place which is subjective and idiosyncratic. The modern perspective from nowhere merges with a place and space that is public, objective, and geometrical. The consumer's landscape stands in an intermediate position, drawing the subjective and objective ends together with its blend of fact and fantasy.

The consumer's landscape draws from each realm of forces elements that are especially problematical in the modern world. Other elements may be more fundamental to particular theories about human behavior, but the elements identified here speak to these theories and also contribute to the geography of everyday life. From the realm of meaning comes the modern awareness that we ourselves—not divine force—are the authors of meaning and thus possess the freedom and the burden to create meaning—a realization that can itself undermine our confidence in meaning. From the realm of social relations, with its increasing bureaucratization and social mobility, comes the awareness, characteristic of modern life, of being in a world of strangers. From the realm of nature comes the force of material objects of the natural world and the fundamental ordering system of space and its tensions between areal differentiation and integration. The world of consumption draws especially on the spatial relations of the natural realm, which can be reinforced by the others (as in territoriality accentuating areal differentiation and integration).[39]

Clearly, the three realms (nature, meaning, and social relations) are dynamically interdependent. For example, spatial segmentation contributes to a world of strangers. A world of strangers makes it difficult to share meanings. Not sharing meaning heightens our awareness of our freedom and burden to create meaning. It also reinforces the distinctions between public and private contexts, which brings us back to the problem of shared meanings in a world of strangers.

These threads are interwoven by consumption, which adds its own dichotomies. How this happens is more closely scrutinized later in the book, for it explains the tensions and contradictions in the consumer's landscape. Among the dichotomies added by mass consumption are that products are *generic* and yet portrayed as *specific,* that they draw some elements of the world *apart* and bring other elements *together,* that they distinguish *self* from *others* and *we* from *they.* Each of these dichotomies can be used to address the tensions in the others; indeed, one dichotomy can activate another. For instance, generic products can be made to seem different by lending products the power to draw things apart or together, thus creating unusual and juxtaposed contexts. A product that places us in unusual contexts has the power to differentiate us from other consumers and help us become our true selves. At the same time, the product itself also becomes more specific and differentiable from other products. These dichotomies and their interconnections weave the threads from the three forces and thus implant modern contradictions in places of consumption.

These mechanisms in the places of consumption operate in a complex and open context: places of consumption, and places of any kind, are rarely pure cases. Even a shopping mall or a Disneyland, though dedicated to creating a consumer's landscape, has to reckon with other processes and functions that become part of that place. The modern tensions and contradictions possessed by places of consumption are often diluted by these additional functions and processes. To be able to examine the tensions of consumption in their purest form would require idealized places of consumption. The closest thing we have to such an idealization is the presentation of products in advertising, the language of consumption. Advertising depicts how products enable us to create places with us at the center. This is why we must pay attention both to the structure of advertising and to actual landscapes. Such an examination will reveal exactly how places of consumption are constructed and how they embody the tensions of modernity.

Expanding the geographical premise into a general relational framework and exploring landscapes of consumption raise questions about the significance of geographical place. The logic of geography must be extended, so that we can assert the self-evident truths that we are in place and that place matters to actions and theory, and so that we can also develop the analytical tools to show how space and place in general—not only places of consumption—do indeed have an effect. An extension of the geographical premise into a framework of forces, perspectives, space, and place increases the power and utility of geographical analysis. The scope of such a framework can be demonstrated by its ability to illuminate the effects of different kinds of places and the effect of places on the moral order. Place, after all, is linked to morality, and places of consumption are continuously evaluated in moral terms.

Place, Morality, and the Consumer's World

There are many ways to think about the connection between place and morality. Places are often described as good or bad in that they promote or diminish the good life, and many of us wish to make our place and our world good or better than it was before. The entire structure of the consumer's world begs to be evaluated in moral terms. Are such landscapes authentic or inauthentic? Are they orienting or disorienting? Does consumption and its construction of place destroy the natural environment, promote social exploitation, and trivialize meaning? Does mass consumption make us free, or irresponsible, agents? In short, do such places have moral or immoral effects?

Ad hoc moral positions justifying or condemning consumption are readily at hand, but it is important for geography to morally evaluate places of consumption, as well as the moral effects of other types of places, through the use of fundamental geographical principles. I argue that an extension of the geographical premise yields precisely such principles, which then can be used to examine any kind of place. This is because moral principles are already implicated in the connections between forces, perspectives, space, and place.

The crux of the argument is that moral agents, especially powerful ones, must be responsible, and that means they must know the consequences of their actions. Most moral theories assume that these consequences, and hence our responsibilities for them and to others, are simple matters, because they consider all agents to be alike and to have limited and circumscribed powers and effects. But this assumption cannot hold in the modern world. As consumers, we are now so enormously powerful that the consequences of our actions are neither immediate nor self-evident. Ideally, they should be worked out in detail for practically each situation. This means tracing, through geographical space and time, the effects of our actions on nature, meaning, and social relations.

Geographical place and space become indispensable means for understanding our effects and responsibilities, because actions travel through space, and place guarantees that we think of the broadest connections among these forces and perspectives. The inside/outside character of place makes us aware of being somewhere, which allows us to see that we are free and morally responsible agents, while it also forces us to view the process from several perspectives, which has the moral virtue of placing us, quite literally, in the position of others and thus able to see the problem from their point of view. Thus to be responsible and moral, we must, at least, have the opportunity to assess, as best we can, the effects of our actions, and see ourselves as able to change our actions. In the absence of a simple and predictable world, and in the presence of incomplete and conflicting theories, such an understanding requires a geographical imagination, using places of various scales. Geographical knowledge becomes, then, a necessary precondition for moral action. A place that prevents us from understanding the consequences of our actions is not moral, even though the consequences of acting in and through that place may be (accidentally) benign.

This moral extension of the geographical premise can form a common ground from which we can evaluate places of consumption. Since these places are disorienting and create the illusion that they are unconnected to other places—that commodities simply appear and disappear from context to context and have no particular history or geography—they place obstacles in the way of evaluating the consequences of our actions. Even though the effects of consumption may not be deleterious, the places themselves prohibit us from knowing this. Thus they promote irresponsibility, which is immoral.

The Organization of the Book

Extending the geographical premise and demonstrating the centrality of places of consumption to modernity are the interlocking and mutually reinforcing aims of this book. Extending the premise leads to a relational geographical framework that will allow a geographical analysis of the consumer's world and of other types of places, as well. An examination of consumption will demonstrate the way such landscapes actually draw together the features of the framework. Part 1 elaborates the premise into a framework. Part 2 proceeds with the geographical analysis of consumption.

The division of the book into two mutually reinforcing parts—the relational framework and the consumer's world—creates unavoidable asymmetries. First, the framework is general enough to apply to numerous other types of places than places of consumption. Still, consumption is stressed, because it is the most accessible and powerful device by which individuals daily create places and transform the world, and because the premise of the book is that being in place affects how we organize and think of the world. Second, the framework attempts to be comprehensive and neutral regarding forces and perspectives, so that place as a concept gives equal emphasis to all of them. If we want to be aware of the consequences of our actions and thus morally responsible, we must focus on all of the forces. Yet, in the analysis of the consumer's world, consumption's effect on meaning may appear to be emphasized more than its effect on nature and social relations. This is because the connection of consumption to nature and social relations has already been examined extensively (by the ecology movement and by social critics, respectively) and because illustrating consumption's reflexive effect on the framework leads naturally to an emphasis on meaning, simply because the framework is a map of intellectual positions.

Apart from this introduction and the afterword, the book consists of seven chapters, organized into three parts. My intention in part 1 is to consider epistemological and ontological issues and their interconnections and to demonstrate how space and place pervade their very conception. Chapter 2 describes the embedding of space and place in the very sense of awareness, which is the basis of abstraction and which requires a movement from somewhere to nowhere. How segmentation of space and place is related to the ability to separate ourselves from the world and to see ourselves as part of the world is considered in the general terms of awareness and in the particular contexts of child development, territoriality, and the historical development of the construction of space. Throughout the discussion, I point to the need for spatial concepts in the very idea of abstraction. Place itself requires us to imagine being both inside and outside. In addition, the movement from somewhere to nowhere, especially along the discursive, scientific axis, has become synonymous with a movement from the private and subjective to the public and objective. These movements along perspectives in turn affect the meanings we give to space and place.

Perspectives also affect how we think of forces, and one particularly important connection centers on the concept of agency. Chapter 3 shows that the sense of ourselves as a force that is not caused to act in a certain way and is free to do otherwise seems to come into focus only from a view from somewhere. This view, which is associated with the "somewhere" of being in place, is necessary to retain if we are to be morally responsible for our actions. This view diminishes in clarity as we shift to more abstract positions that find reasons or causes for actions. These reasons and causes (and reasons often become causes) make the free agent virtually disappear and the sense that we could do otherwise into an epiphenomenon. From these more distant perspectives, our actions still instantiate forces in the structurationist sense, but we are not morally responsible for them because we may be acting as vehicles of these forces.

Awareness, though, does not happen all by itself. It requires other forces to propel it, which then leads us to the issue of forces directly. In chapter 4, I examine claims about the primacy of the forces of nature, meaning, and social relations and the intellectual deadlock among these claims. Moreover, I argue that, though space forms part of the ontology of each of these realms, the fullest integration of space and place and of these realms results from geographical analysis. Space and place provide a means of integrating perspectives and forces.

Part 2 examines the kinds of places that consumption creates. Chapter 5 explores aspects of modern culture and places that are somewhat independent of consumption: utopias, national territory and national monuments, and places created by and portrayed in television. All three integrate forces and perspectives, but they do not cause us to create contexts in day-to-day activities, as consumption does. Consumption engages virtually everyone; consumption creates and transforms places; consumption combines particularly modern facets of forces and perspectives. Landscapes of consumption, then, create and recreate the tensions of modern life, which is why they deserve our attention.

Chapter 6 presents a geographical model of consumption in the form of a loom that fits within the relational framework. The model demonstrates that consumption is a place-creating device, that its idealized form is in advertising, and that it contains a structure that enables consumers to draw together specific elements of forces and perspectives to create places that are in a fundamental sense out of place. These disorienting qualities of place are precisely those of modernity and postmodernity, and not surprisingly, the places of consumption exacerbate these tensions and contradictions.

Chapter 7 demonstrates how these tensions identified in the model actually appear on the ground in real geographical places. Department stores, malls, shopping strips, revitalized downtowns, tourist places, homes, and theme parks all contain these contradictions and assume the role and appearance of real-life advertisements. Since these places are out of place and disorienting, and since so much of our time is spent in them, they perpetuate the modern contradictions. Such contradictions are not simply intellectual concerns, though they do affect the way we conceive of forces and perspectives; they involve, as well, actions that transform these forces and perspectives.

How, then, can we evaluate consumption? This is the particular question addressed in the part 3. Chapter 8 reconsiders the relational geographical framework in moral terms. Forces and perspectives are loaded with moral content, and moral issues depend on space and place as much as does the general framework. This is shown first in the way that problems of equality, freedom, and responsibility cannot be conceived of without geographical space and, second, in the way moral positions range from local to global perspectives. It is then shown in the way these ideas enter into a discussion of paradise and utopia. Geographical knowledge and imagination are necessary for responsible actions, which are prerequisites for morality. This is a general geographical condition of morality, and it is violated by the consumer's world. Equally important, this geographical basis of morality can be used to evaluate other types of places and actions.

Finally, a note about the illustrations. They are collected in a gallery at the center of the book. Figure 1 provides a general conceptual map for the relational framework. Figure 2 is a reworking of the relational framework by the particular geographical activity of consumption. The rest of the illustrations show how consumption is advertised. Presenting them together allows not only easy reference but also a visual portrayal of the development of the themes in the book that complements the verbal exposition in the text. The advertisements were selected to serve both as illustrations of the points in the text and as a pictorial narrative. Although the advertisements are referred to in the text as illustrations of particular tensions of consumption, most contain other tensions as well.

I

The Relational Framework

· 2 ·
Perspectives from Somewhere to Nowhere

The effects of a consumer's landscape result from both its distinctive and its general spatial qualities. This chapter is about the general properties of space and place and their centrality to experience and action. The geographical premise is elaborated here into a relational framework that reveals how geographical space and place draw upon and help constitute the primary intellectual forces and perspectives in modern debates. Although forces and perspectives are interrelated, this chapter concentrates on perspectives.

Numerous views or perspectives allow us to see the world in geographically interesting ways, and many of these have been the object of rigorous analysis.[1] I focus here on those perspectives that lead from somewhere to nowhere and that have the most to say about modern place. This means concentrating on the discursive, scientific axis and on the tendency for modern culture to consider the view from somewhere as a more personal, subjective, and idiosyncratic perspective and the view from nowhere as a more public and objective perspective. The critical point is that the very awareness of being in place provides the momentum to propel our views along the path from somewhere to nowhere. This momentum is supplemented by other forces, which are also addressed in this chapter. The discussion begins with a description of how meaning assists movement from somewhere to nowhere, for the distinctions between somewhere and nowhere first emerge in the process of awareness.

Personal Place: A View from Somewhere

Our experience of being alive is inextricably tied to our sense of being in place. This is what is meant by the *personal sense of place*.[2] Place, here is not simply a location in space that all things occupy because they exist; that meaning of *place* as location comes into clearest focus when we see ourselves from the outside, though it is intimated even in the personal sense of place. By being in place, I mean the more personal sense of place that stems from the fact that we, ourselves, are the locus of sensations and actions simply by virtue of being in the world. The term *world,* with its Germanic stem *wer,* means "the earthly state of humankind" ; in other words, it is equivalent to *being*. But being is also active, and our actions and consciousness construct and sustain our personal sense of place. Everything we experience and every action we undertake happens in place. As we move, so does our personal sense of place. It is created from the interweaving of our everyday thoughts and actions with those of others. As our attention and actions wax and wane, so does place. Place, then, need not correspond to a fixed extent in space or to a specific area. This is the meaning of place that humanistic geographers have attempted to isolate by shunning the view from nowhere and inserting themselves introspectively and empathetically into the experience of being in place.[3] Though the content and development of our personal sense of place depend on our position in our culture, place, and historic period, this sense of place is assumed to be shared by all.

Irrespective of its contents, this personal sense of place has two contradictory yet interrelated sides. As a locus of experience, personal place provides a holistic sense, interweaving elements from the realms of nature, meaning, and social relations. From its other side, it sets in motion the potential to see the world from somewhere else, and then unravel the threads and trace them back to the particular realms. The key to understanding the relation between these moments lies in the connection of place to awareness. The holistic sense of place occurs at the preconscious level, of ourselves as the locus of experience and yet not aware of that fact; whereas, at the very same time that we are beings in place, our potential to become aware of this fact can be activated, which initiates the view of ourselves in place as seen from outside ourselves somewhere else.

Personal place, then, contains both a preconscious and a conscious side. (*Preconscious* is used instead of *subconscious* to avoid the latter's association with psychoanalysis. Preconscious here refers to awareness of place and not to awareness in general.) To possess the concept of place, we must be aware, and once we are aware, the mind—not alone, but in conjunction with the force of social relations—unravels the weave of personal place and sets in motion the movement from somewhere to nowhere.[4] We can also move from the conscious back to the preconscious.

I begin with the preconscious and then consider its potential to unravel and transport us to a view of ourselves from outside our place.

Preconsciousness

Suppose we are comfortably settled in our chairs at a table in a restaurant, relaxing over coffee. We have nothing particular in mind, our thoughts are drifting. Suddenly something changes. We become alert, trying to sense what it can be, and in a split second realize that the lights have been dimmed, signaling the shift from the afternoon luncheon period to the more atmospheric dinner hour. The change in lighting made a change in place; it darkened it. But this change also points out how almost anything else could have changed the place. A sound, a smell, a touch, or a thought could have drawn us up.

This particular change is not an important one and can soon be assimilated, so that once again we settle back and let our minds wander. But change can be more drastic and alter the quality of the place. If other patrons are inconsiderate of the staff or of each other, the place can become unpleasant and affect how we feel and act. We may no longer be able to concentrate or to engage in conversation; we may even feel threatened. Or the restaurant's service may deteriorate because the surrounding streets need repair and food supplies cannot be delivered. The neighborhood may be declining, so that the place has become dangerous. More drastic changes could occur: an earthquake, a tornado, or a terrorist bomb blast could literally shatter the place and make our world crumble. The changes, though, could also come from within ourselves. A tragic memory or a horrible fear could darken our world and make the place feel uncomfortable. The same might happen if we are stricken with a sudden wrenching pain: the grip of pain and fear may be so tight that our world shrinks to the point of excluding all else.

Personal place depends on and influences actions. Virtually anything can help constitute this place, and virtually anything can change it. The changes can be local or global: the place can shrink, or it can expand, even to include the entire universe, if we happen to be thinking cosmically. Personal place does not simply exist—it requires constant (though often preconscious) effort to support and is connected to the more physically visible and stable places in the landscape. In this example, the personal sense of place occurs in a restaurant, which is a maze of planning and activity, from the purchase, preparation, and serving of the food to the planning and maintenance of the physical structures. And its patrons, as well as its employees, rely on webs of interdependencies that can extend over the entire globe.[5]

Until the restaurant's environment changes, we are unaware of being in place. This does not mean, though, that we are not experiencing it, but rather that we simply are not conscious of it (although we are conscious of other things). I call this the preconscious part of personal place. If personal place is fixed in location and extent, then the preconscious part would be close to a sense of rootedness. Much of our lives are lived in this state, and a great deal of personal and social energy is expended in maintaining it—because the preconscious part of personal place sustains routine. Parts of space become structured by everyday actions, and these structures support the actions and ways of daily life. In this way, daily activities in space unconsciously reproduce the particular forces that have helped determine these routines. They become "geodoxies." Under certain circumstances, the slightest alteration of these spatial routines can be interpreted as resistance to authority.[6]

Spatial routinization, though its particular forms instantiate and even disguise aspects of authority, can also be liberating, allowing us to concentrate on other things. If the places we occupy do not work smoothly—if the roof leaks, the lights go out, and the furnace fails—we become aware of place, but we are not able to attend to much else. Consider how many different ways our society, and we individually, construct and maintain places so that we can become unaware of them. The restaurant is one. The tables and chairs are spaced so that customers can attend to their meals and conversations without intrusion and waiters can move about efficiently and along predictable routes. Internal physical conditions are comfortable in all seasons and times of day. The restaurant's kitchen is laid out to encourage efficiency and routine and is stocked with food. The restaurant itself is accessible to customers. And even as customers we make an effort to create a place that we can then ignore, moving our chairs a comfortable distance from the table, readjusting the position of our plates, cups, or silverware.

In the workplace, the arrangement of the material environment might be even more fine tuned so that it does not intrude on our work. Time and motion studies may stipulate the arrangement and position of materials, tools, and the position and orientation of our bodies so that we can work at optimal efficiency. It is also true of our journey to and from work. It would be a taxing, and perhaps impossible, journey if we constantly had to remember where we were and where we were going and if society did not assist in virtually every detail of navigation by arranging the location and form of roads, with their traffic lanes, road signs, traffic lights, and so on. These signs, signals, and spatial configurations work so smoothly that, once we know our way between work and home, we might not need to concentrate on our journey at all.

This rootedness in place, even a rootedness that can be carried or transported, is the geographical backbone of personal and social routine.[7] Without it, we would continuously need to construct our world anew, and we could be paralyzed with indecision. The sense of rootedness frees us for other things, things that must also be place bound. Insofar as our lives must be based on routines, this preconscious sense of personal place both nurtures and frees us. It sustains actions and allows new ones to develop. Yet if these routines are somehow limiting or unjust, routinization of action at a preconscious level can help perpetuate bad habits and injustices.

This preconscious sense of personal place is complex and integrating. Even though its contents can vary from context to context, and are brought to light when the system changes, or does not work properly, or is resisted—this preconscious part of a sense of place has a common structure. Yi-Fu Tuan shows how preconscious spatial activities contain polarities of meaning reflecting our shared biological asymmetries. For instance, all nonhandicapped humans walk erect, have a front and a back, and most are right-handed. These biological asymmetries are imparted to our particular orientations in place. We are, for the most part, more comfortable with the world ahead and in front of us than with what is behind us; we generally favor our right side over our left; and our upright position makes us impart positive qualities to height: *taller, higher,* and *above* are usually "better" than *shorter, lower,* and *beneath.* Such associations do have numerous exceptions (e.g., although we usually wish to soar to the highest level, we also value qualities that are deep and profound), and they do not consistently function as basic organizing principles in the experience of place in our modern complex society.[8] Rather than explore the preconscious level further, I move in the other direction and consider the means by which the personal sense of place is linked to awareness of place and how this awareness, in conjunction with specific forces, leads to an unraveling of the holistic sense of place and the development of different perspectives.

Awareness

Consider first the development of perspective. Changes in the environment or in ourselves make us aware of being in place. No longer are we simply the locus of experience; we also possess a perspective that helps us to see that we *are* this locus. We have to some extent stepped outside ourselves to attain a perspective that contains our original position as a particular perspective.[9] We no longer simply react to the events in the restaurant—we also see that we *are* reacting in the restaurant. This new view is more comprehensive and objective, because it relies less on our personal perspective and attains a perspective that others share: objectivity in this regard is based on intersubjectivity. According to Thomas Nagel, objectivity must "point beyond mere intersubjective agreements" to real concrete experiences that can be shared.[10] This means that, in addition to intersubjectivity, objectivity requires a realist assumption that there is indeed a reality that we have in common. This move to outside of ourselves can never be complete, but it can become more distant or abstract and, through intersubjectivity, communicated and shared, so that we can imagine ourselves to be simply a location or point in space, surrounded by other objects and forces. The same imaginative movement can also conceptually remove us from the place by having us imagine that we are no longer there and view the qualities of the place that we once occupied as simply bundles of intersecting forces in a spatial system. Of course, at any moment we can imaginatively reinsert ourselves in our own, or any other, place and see how it feels from the inside, making it once again a personal place, if only in our minds.

This imagined movement from somewhere (within place) to somewhere else (outside and more distant) is a general feature of awareness and abstraction (and is suggested by the cone in figure 1). In our culture, with its high degree of specialization, this movement can occur along several well-developed lines of conceptualization (suggested by the axes in figure 1), including the discursive and scientific, the aesthetic, and the ethical. Each axis draws different facets of space and place into focus, at different levels of generalization. These lines of conceptualization are components of our daily experiences, and their different qualities reveal facets of our landscapes that are important in the contemporary world.[11] But our culture places the greatest emphasis on the axis that pertains to the practical aspects of experience and that uses discursive forms of representation, from ordinary language to the highly specialized terms of science. (Ordinary language can also be thought of as the basic nonspecialized form of presentation for the other modes.) The imagined movement from somewhere to nowhere along this particular axis is especially important here, because the abstract end has come to mean the public, objective view, where space is seen as geometrical and place becomes a location in that space. This leaves behind the view of somewhere and makes it seem a personal, subjective, and even idiosyncratic perspective. The movement from somewhere to nowhere, and the axes along which this movement takes place, are borne of the general impulse to transcend ourselves (to paraphrase Nagel), but its qualities are molded by the forces from the realms of nature, meaning, and social relations. Yet the very movement along these axes affects how we conceive of these forces and their connections to place. Indeed, it makes these forces and realms more distinct.

The personal sense of place, especially for the discursive and scientific axis, comes into sharpest focus only with a view from somewhere. This sense of place is able to integrate so many elements that only a few are ever disclosed, and those few come to light either because something in the place changed or because we move away from this view and examine the place from somewhere else. As we move along the path to nowhere, the holistic quality of this sense of place unravels until we see it as containing forces that may not be causally connected. The view from nowhere tends to conceptually separate these forces into events that could theoretically occur elsewhere in space. By distancing us from our experience as the locus of integration, the view from nowhere must rely on theoretical knowledge to provide connections. And even if it succeeds in developing such knowledge, the view from nowhere presents a picture wherein the integration of such forces occurs anywhere in space that conditions permit, and there is then no necessary link between events and place.[12] Moving from somewhere to nowhere weakens the bond between events and place, and as these drift apart and their relation becomes more contingent, our view from nowhere focuses instead on space and distance. These become the major variables in a public discourse.

The views from somewhere and nowhere along the axes are not complete opposites or even complete by themselves. Each is only partial and contains elements of the other. The view from somewhere draws attention to the fact that we are in place, and this awareness alone begins our movement to somewhere else. The view from nowhere can never be attained completely, because at the very instant we strive to conceive of it, we must still be somewhere, which in turn affects our view.[13] This movement from somewhere to nowhere, especially along the discursive and scientific axis, draws into focus many of the meanings of contemporary geographical space and place and how they become intertwined with specific forces. Our dynamic social organizations find it difficult to focus on and share the contents of the personal sense of place, which is why it seems private, subjective, and idiosyncratic. Our need to quickly and accurately share a perspective makes the view from nowhere an excellent candidate for the public and objective view, so this view becomes embedded in the look and feel of the landscape: in its grid patterns, its regular and rectangular architectural structures, and the experience of travel on railroads and airplanes as duration through time.[14]

This public, objective, landscape is tied to the personal place because we must inhabit it. And these personal senses of place can be quite different in content than their surrounding public senses of place and space.

As passengers in a plane hurtling through empty, undifferentiated, space, we could still be in an extremely rich personal place within the cabin that contains specific environmental conditions, social relations, and our own thoughts and feelings.

I have discussed how our impulse to transcend ourselves leads us along the paths from somewhere to nowhere and how this movement, especially along the discursive and scientific axis, is conjoined in our society with the categories of objective and public, on the one hand, and subjective and private, on the other. This conjunction is so strong that the direction toward nowhere along these axes (but mostly along the discursive and scientific axis) signals a movement toward a relatively more public and objective position, while the direction toward somewhere points toward a relatively private and subjective view. I thus use *objective, public,* and *nowhere* interchangeably and *subjective, private,* and *somewhere* interchangeably. The role of place in these associations is not simply metaphorical—it is virtually impossible to imagine being aware without thinking geographically: as beings, we are always in place.

This basic relation between place and awareness is supported by the fact that, even in philosophy (which we might think would not need to take the sense of place seriously), the role of place and the distinctions between somewhere and nowhere are essential to the analysis of epistemology. Viewing the world from afar to attain a certain critical distance is a basic feature of human knowledge. As Martin Buber put it: "It is the peculiarity of human life that here and here alone a being has arisen from the whole endowed and entitled to detach the whole from himself as a world and to make it opposite himself." To Buber, this movement is the basis of objectivity, "which is nothing more than the act of making things (the world) distant from oneself."[15] Thomas Nagel draws attention to the same impulse but makes it more of a continuum, linking somewhere with subjectivity and nowhere with objectivity. The philosophical problem, according to Nagel, is

> how to combine the perspective of a particular person inside the world with an objective view of that same world, the person and his viewpoint included. . . . [This problem] faces every creature with the impulse and the capacity to transcend its particular point of view and to conceive of the world as a whole. . . . The difficulty of reconciling the two standpoints arises in the conduct of life as well as in thought. It is the most fundamental issue about morality, knowledge, freedom, the self, and the relation of the mind to the physical world. . . . The process is one of degree and leads to greater and greater objectivity. . . . To acquire a more objective understanding of some aspect of life or the world, we step back from our initial view of it and

> form a new conception which has that view and its relation to the world as its object. In other words, we place ourselves in the world that is to be understood. The old view then comes to be regarded as an appearance, more subjective than the new view, and correctable or confirmable by reference to it. The process can be repeated, yielding a still more objective conception. . . . A view . . . is more objective than another if it relies less on the specifics of the individual's makeup and position in the world, or on the character of the particular type of creature he is.[16]

However, objectivity can never be complete. It must always incorporate the other part of oneself that is the subject. "The most objective view we can achieve will have to rest on an unexamined subjective base. . . . We can never [completely] abandon our point of view . . . we can only alter it."[17] Though the movement from somewhere to nowhere is borne in awareness, it is not simply a matter of the mind. It requires sustained individual and social effort to expand and elaborate these perspectives, and each society accomplishes this differently and to different extents. In Nagel's terms, "the objective self is only part of the point of view of an ordinary person, and its objectivity is developed to different degrees in different persons and at different stages of life and civilization."[18]

The development of the objective point of view depends to a considerable extent on actions that involve spatial manipulations. I discuss this in two areas: the development of awareness in the child and society's use of territoriality as an instrument of control. In both cases, manipulation and conceptualization reinforce each other, their conjoint effect is to enable the subject (either the child or agents of society) to live in a more objective world—to move up, as it were, the axes, but especially the discursive and scientific axis.

Space, Place, and the Child

Child development literature stresses how important our spatial manipulation of our surroundings is to the attainment of a mature awareness of ourselves in the world. The child begins life without distinguishing between subjective and objective, somewhere and nowhere. "All that we are fully justified in assuming for the mentality of the newborn child is a blurred state of consciousness in which sensorial and emotional phenomena are inseparably fused."[19] Intellectual development is the progression from a global and syncretic view, in which child and world are virtually fused, to a discrete, differentiated, articulated, and hierarchically integrated view, in which the child sees itself as a subject and also as an object in the world. This awareness of ourselves in place (and the movement in awareness along the axis from somewhere to nowhere) is an outcome of the child's manipulation of objects in space and a development of a conception of space.

Before considering how spatial interaction and conception is intimately involved in intellectual development I will discuss the two primary stages of intellectual development as articulated by Jean Piaget. The first involves a perceptual separation of self from the world, culminating in our ability to see ourselves as objects in a stable world. This development begins in the first months after birth and is well established by the end of the second year. By this time, the child has developed a relatively clear and stable perception of its immediate surroundings and of itself both as a being apart from the rest of the world and as one physical object among others in place in the world. Although from this stage on, we distinguish ourselves from the world, objective and subjective facets are still closely connected and are often intertwined in the process of evaluating self and world.

The second stage is the development of conceptual or symbolic thought; its beginnings overlap the end of the first stage. Conceptual or symbolic thought is the ability not simply to perceive but to represent ourselves and the world through thoughts and symbols. Symbols can be defined as anything—words, thoughts, pictures, gestures, or objects—that stand for, or represent, something else and that do so even when that something else is not present. (This last condition differentiates symbols from signs, which point to things that are physically there, and from signals, which are triggered by something else.)[20] Because symbols are the devices by which we can represent the world, they provide us with a means of transporting ourselves (along the axes) from somewhere to nowhere. They thereby help to drive a deeper wedge between the subjective and the objective.[21]

The development of the ability to conceptualize and use symbols becomes more abstract until, as young adults, we perceive—and conceive of—ourselves as objects in a stable world. In other words, we are able to imagine ourselves from an almost infinite set of places and perspectives. Beyond this point, facility with conceptualization continues, but along the lines set out by particular cultures. In our society, this means a tendency to specialize in particular symbol systems or modes of thought, such as science and art.

The shift in awareness from perception to conception and its movement toward abstraction does not just happen; it involves the continuous manipulation of objects in space and of particular spatial relations. The child first senses objects as if they were simply events that appear and disappear without relation to one another or to the child. This disconnected sense of an external world can be described geometrically as topological space—or more accurately, as a set of places composed solely of objects and their shapes, but without accounting for their relative positions in space or for their positions vis-à-vis the child. Thus the child would not know that a more distant object seems smaller than a nearer one or that an object that moves can be expected to continue along its trajectory even when it is no longer in sight. Such knowledge is acquired through the continuous movement and manipulation of objects (including the self) in space.[22]

As the child becomes aware that objects remain in the world even when they are not in view, that the child is part of this world, and that the objects are certain distances from each other and from the child itself, the child moves from a purely topographical sense of space composed of discrete places to a projective geometrical sense of space composed of interrelated places. Such a space allows for the permanence of objects, their relative positions, distances, and movements in space, but it considers space from only one perspective: the child's perspective at the center of an egocentric place. The child still cannot see itself from the outside and thus possesses only the preconscious sense of personal place.

By the second year, the child develops a sense of permanent objects that move, a sense of a spatial system that contains them, and the capacity to see itself from the outside, as an object in that space. This means the addition of conscious layers to the preconscious sense of personal place, which is accompanied by a shift from an egocentric projective geometry to a sense of Euclidean space. Euclidean geometry is a composite of projective geometries, containing the particular perspective of each projection. And just at the point when the child's perception matures, it develops conceptual capacities or symbolic operations that recapitulate the stages of perceptual awareness and build upon the sensory-motor experience. The first stage beyond the sensory motor is the preoperational (at two to seven years of age), in which conceptual thought seems to be closely tied with action. At the beginning of this period, the child conceives of objects in an egocentric way, in terms of remembering or picturing its physical relations with them; and such images of previously manipulated or perceived objects are often uncoordinated and short-lived. Although the child perceives a spatial manifold, it does not conceive of a single spatial continuum, nor does it think that objects in space maintain their shapes and positions independent of the child's viewpoint. The child's conception of spatial relations is primarily topological.

In the second stage of symbolic operations, called the concrete operational (at seven to ten years of age), the child notices that space has invariant properties independent of the objects in it. At this stage, the child is mastering a projective conception of space. It now sees that the distance between two objects remains the same when a third object is placed between them, and that the winding path between two points is longer than a straight path.

The last stage is formal operations (at eleven to fifteen years of age). At this level, the child is able to think about thought itself and to consider possibilities in addition to actualities. At the end of this stage, the child attains a mature conception of space as a three-dimensional manifold, the spatial properties of which appear stable and Euclidean and provide a framework for possible as well as actual events. Even at this point, thought is still internalized action. The interdependence of spatial action and awareness is made clear by Piaget:

> Spatial concepts are internalized actions, and not merely mental images of external things or events—or even images of the results of actions. Spatial concepts can only effectively predict these results by becoming active themselves, by operating on physical objects, and not simply by evolving memory images of them. To arrange objects mentally is not merely to imagine a series of things already set in order, nor even to imagine the action of arranging them. It means arranging the series, just as positively and actively as if the action were physical, but performing the action internally on symbolic objects.[23]

The young adult finally attains a conception of a stable Euclidean world, which provides her or him with a perspective from somewhere *and* nowhere. This kind of geometry, moreover, is an ideal foundation for a shared public perspective, because once mastered, it allows others to attain perspectives that they do not literally possess. The ease with which it provides common vantage points recommends this geometry as an appropriate public description of space and place for a dynamic culture and its world of strangers. Of course, Euclidean geometry is not thereby the most sophisticated geometry; it is simply a useful one for describing how we attain an awareness of the world. And when I say that we see the world in a Euclidean framework, this does not mean we are aware that such a world is Euclidean (or projective or topological, for that matter). Such awareness comes from being tutored in geometry. Rather, it means that what we see and how we think can be described as though we possess (or are constrained and enabled by) such a geometrical structure.

The discussion has emphasized how the development of awareness of self and world involves the continuous manipulation of objects and materials in place and their spatial relations. Indeed, even the most abstract mental operations still internalize basic psychomotor involvements with objects, space, and place, and traces of these operations are retained in the very symbols themselves. For example, the spatial asymmetries of our own bodies (i.e., front/back, up/down), along with their emotional imports, become embedded in our prepositions, prefixes, and other parts of our language and actually affect the speed with which we recall and understand them. But, as I have said before, this developmental process is also a product of particular forces of social relations and meaning. In the Western context, these have driven the Euclidean sense of space to yet a more abstract plane by emphasizing its geometrical qualities. And, as a geometry, this space appears to exist independently of the activities and things that helped constitute it: it denies its own history.[24] This capacity of Euclidean space—to appear to have a structure independent of substance and thus to be conceived of absolutely and nonrelationally—requires another important spatial component, and that is the capacity to control space territorially. Controlling space permits places to form and allows space to be literally emptied and filled.

Territoriality and the Production of Space

A geographical truism is that events and processes are unevenly distributed over the surface of the earth. This unevenness is called *areal differentiation* and is one of the principle threads that consumption draws from the realm of nature.[25] Geographically distinct places such as deserts, savannas, and tropical rain forests are products of natural processes (though recently requiring human assistance) that permit things to occur in one area and not in another. Human activity is also unevenly distributed over the earth, but in the human world most areal differentiation does not just happen and then remain that way (which is the picture presented by some of geography's location theories) but must be formed, maintained, or changed, by human effort.[26] A cultivated field is a place, but it is also a human construction and requires such efforts as plowing, fertilizing, and weeding to maintain it. Even more basic, such a place exists because decisions have been made that all else but certain specific processes and events will be excluded from that area: trees, bushes, and weeds will not be permitted there, nor will animals or even most humans. If this field were not tended, natural forces would eventually make it another kind of place entirely.

The same holds true for virtually every other place in the human world. The cities we inhabit, the rooms we live in, the roads we drive along exist because we make an effort to have them contain only some things while excluding others. When we control areas of space, we create territories. Territoriality—the act of creating territories—is the geographical exercise of power. Territoriality is a strategy for maintaining social order and imparting meaning to phenomena. It is possible, though not always feasible, to allow things to occur and interact in space without our controlling the area or clearing a place for only certain kinds of things, social relations, and meanings to take place. But in a complex society, where many things and processes compete for space, most of our social activities and the meanings we create would not exist without territorial control. Without territoriality, although we would still possess a personal sense of place, this personal place could not be inserted into any other kind of place, such as rooms, homes, restaurants, farms, and cities. These other places are the backdrop of society. Without them, we would expend all of our energies creating order and meaning for even the most elementary activities.

Territoriality creates and sustains real places that provide us with support. It also plays an important role in the development of our subjective and objective senses. First, territoriality provides us with the places in which we can imagine ourselves to be and thus allows us to think of ourselves as from somewhere else. In real life, the components of Euclidean space are not really coordinates or locations in that space but, rather, places. And these places that we imagine going to and from—a room, a home, a school—are sustained by territorial decisions that assure that only specific kinds of functions and meanings occur in that place.

Second, territoriality allows us to think of space as a container for the spatial properties of events. The influence and authority of a city, although widespread, is legally within only its political boundary. When the social relations and meanings to be contained by the territory are not present, the territory can be thought of as conceptually empty. Territoriality, in fact, helps create the idea of emptiable place. Take a parcel of vacant land in the city. It is described as an empty lot, though it is not physically void, for there may be grass and soil on it. It is empty because it is devoid of socially or economically valuable artifacts and meanings. By creating places as containers, and by emptying and filling them, territoriality helps us think of space abstractly and as though its existence is independent of events. The capacity to empty space reinforces its role in the creation of a public, objective view.

The movement from somewhere (subjective) to nowhere (objective) is the product of awareness, as awareness develops through the manipulation of objects and spatial relations. This awareness is fostered by the existence of places that are socially constructed through territorial control. Curiously, this move toward the objective promotes a heightened sense of self and personal space and place. Personal place is both preconscious and aware; awareness is possible only because a person can be conceptually somewhere and somewhere else at the same time. The tension thus created makes for self-awareness and reflexivity, a growing sense of our individuality and the individuality of our particular place. I now show how this turn toward individuality and self-awareness happened historically in Europe, as Europeans moved up the discursive and scientific axis. I then show how this move resulted in quantification and geometrical space. It is a principal purpose of this book to demonstrate that both the inward turn to individuality and reflexivity and the outward turn toward objective reality have enormous consequences for what we know as the modern world and, in particular, the world of consumption.

Segmented Worlds and Self

Yi-Fu Tuan provides a historical and geographical analysis of the joint development of the idea of self and the segmentation of space—a segmentation that is often created and sustained by territoriality.[27] Tuan argues that to raise the question, Who am I? "Presupposes an ability [literally] to stand apart from the group." The young child may eat on its mother's lap, the older child may eat sitting next to the parent, and the adolescent child, beginning to develop a greater sense of self, eats farther apart, even alone. In most societies, people want to be able to withdraw from others for certain biological needs, such as defecation, sleep, and copulation. "These are times when men and women feel vulnerable. The sense of vulnerability implies an awareness of self,"[28] not only because we are vulnerable biologically, or off guard, but because we are not comporting ourselves as we think we should—not meeting our standards, and thus experiencing shame. Withdrawal requires that there be distinct places we can retreat to, places necessary for the development of self.

While all societies possess some spatial segmentation and some sense of self, the differences in degree are enormous. Nonliterate societies have the simplest technologies, with the fewest internal territorial divisions; a few even live without purposely built shelters. Individuals in these cultures also exhibit little sense of self apart from the group. "In nonliterate culture," according to Tuan, "the boundary between self and society is much less sharply drawn than it is in complex urban civilizations and particularly in Western civilization."[29] Comparing traditional African societies to our own, Dominique Zahan writes: "To define the self, we separate it from the other, whereas in Africa the opposite is the rule."[30] And Tuan adds, "the African tends to define himself by that which he receives from others at any moment. [The] self is more social than [the] individual."[31]

Similar examples can be found in other preliterate societies. In the Archaic age of Greece, for example, the self, like the self of many primitive peoples, lacks clear subjectivity. The heroes of the *Iliad* do not seem to believe that they initiate their own thoughts. To them, their thoughts are external and often divinely inspired. Their dreams are not internal dialogues between elements of their subconscious but rather conversations with deities. Even their heroic acts are attributed to external forces.

In the Greek Classical period (which was literate), there was much greater elaboration of the individual, but still he or she was secondary to the social entity—which in this case was the group defined by the *polis*. "In its classical form, the polis required the action—the initiating act—of free citizens. Individuals, endowed with rights, asserted themselves. But they did so always for the common good. Despite gains in the idea of the individual worth, what truly counted was the social whole. In classical thought from Plato to Cicero 'the whole is prior to the parts, is better than the parts, and is that for the sake of which the parts are, and wherein they find the meaning of their existence.'"[32]

In the Medieval and Renaissance worlds, the correspondence of the sense of self with spatial segmentation were closer to our own. "In the early Middle Ages, the house, whether that of a lord or peasant, was basically a barnlike structure with a central hearth and a roof open to the ceiling."[33] In such a world, there was little privacy and correspondingly little sense of self. People thought of themselves principally as part of a household, which had an economic base and included family members, servants, and retainers. In the course of time, rooms were added to this central room, known as the hall, but these other rooms were not necessarily set aside for particular functions. Rather a room could serve as a place to eat, or a place to gather, or a place to sleep, depending on the time of day. Imagine the chaos, writes Phillipe Aries, that "reigned in these rooms where nobody could be alone, which one had to cross to reach any of the connecting rooms, where several couples and several groups of boys and girls slept together (not to speak of servants, of whom at least some must have slept beside their masters), in which people forgathered to have their meals, to receive their friends or clients, and sometimes to give alms to beggars."[34]

It was not until the Renaissance that most of the larger houses came to have rooms set aside for separate functions, and yet these rooms were still interconnecting. Hallways and corridors designated solely for traffic and for entering and leaving rooms were not introduced until the commercial revolution of the seventeenth century, and then only the rich could afford these additions. Later, even modest houses had corridors and specialized rooms. It is significant that the introduction of hallways and the specialization of rooms was accompanied by a corresponding change in the city, which saw increased segmentation: specialization of shops and the clearing of city streets for the sole purpose of transportation.[35] These increases in spatial segmentation were accompanied by a separation of the individual from the group and a deeper and more elaborate sense of self.

The fourteenth and fifteenth centuries saw the rise of smaller family units.[36] By the eighteenth and nineteenth centuries, the family became spatially segmented within the home, and the home became ever more separated from places of work, education, and recreation. With the development of these specialized places, the use of the word *I* occurred with greater frequency in literature. "Words such as 'self-love,' 'self-knowledge,' 'self-pity,' 'ego,' 'character,' 'conscience,' 'melancholy,' and 'embarrassment' were to find their way into English and French literature and were used in the modern sense."[37] Mirrors, too, became commonplace, allowing people to see themselves as part of the world out there.

Along with this distancing of self from a segmented world evolved a fragmentation of the senses, with a reliance on sight. We have come to expect to see buildings, streets, and cities and not to hear, to smell, or even to touch them. As Tuan points out, reliance on sight makes places appear more distant and less enveloping, for sight is the most distant and analytical sense of all. "In an earlier age when people were less infatuated with sight and less rationalistic, environment probably seemed more fluid than it does now. Nothing then was strictly delimited. Material objects did not preserve their separateness and identity."[38]

Historically, then, there was a correspondence between increased territorial segmentation of the external world and an increased sense of self apart from the world. The net effect of a highly defined self separated from the world is ambivalent: it can reward us with a "sense of independence, of an untrammeled freedom to ask questions and explore, of being clear-eyed, without illusion, rational, and personally responsible," or it can lead to feelings of "isolation, loneliness, a sense of disengagement, a loss of natural vitality and of innocent pleasure in the givenness of the world, and a feeling of burden because reality has no meaning other than what a person chooses to impart to it."[39] This freedom and burden to create meaning is one of the important components of modern life drawn upon by the consumer's world.

Public, Objective Space: A View from Nowhere

Although territorial segmentation leads to an increasingly complex sense of self and to personal and private places, this and other social and intellectual forces also move us in the opposite direction, to a view from nowhere: a view that is public, objective, and geometrical. Rather than discuss the general history of the shift toward an abstract geometrical sense of public space, parts of which have been addressed elsewhere,[40] I draw from this the more particular issues of the relations among quantification, abstraction, time, space, and territoriality as they help create an abstract empty space. These are focused on below in the context of capitalism, which is as much about one force as about another.[41]

Quantification

In the premodern Western world, problems of fact, of economic worth, and of right and wrong were resolved mostly by custom and religious tradition. Modernization supplanted this world with a dynamic social system containing a plurality of beliefs and perspectives. Human experience, in the form of empirical knowledge and sensation, replaced scripture, tradition, and authority as the foundation of knowledge. Humans took center stage. But these were humans who possessed diverse customs, values, and needs and who now had to be integrated within an ever-larger social system. How could they agree on the significance of things? They needed a readily accessible perspective from which to view the world, and the application of quantification to fact and value helped to form this common perspective. In the economic realm, value was no longer based on use and custom but became measurable in terms of money. The market mechanism of supply and demand provided a powerful device for expressing this measure. As market activity expanded, things became valued in terms of their market price rather than their traditional value or usefulness, which allowed people from different cultures and backgrounds to speak the same economic language. Money, as a "mediator and regulator" of social relations provided a means of integrating a world of strangers.[42]

Quantification through money expressed what was of economic value. Quantification also helped specify the facts of the world. No longer did people look to the Bible or other types of authority as the fundamental description of reality. Science emphasized human experience, and what counted as fact was the frequency of agreement regarding observations. Facts were described in terms of such quantifiable units as location, size, and weight, just as goods were described in terms of price. Quantification even penetrated morality in the form of mass politics and voting (which allows a count of hands to decide what ought to be done) and also in universalistic ethical equations, such as the greatest good for the greatest number as a determination of what is just or unjust. Experience of all kinds thus became more amenable to measure. Quantitative terms to describe empirical reality and economic value were more useful to a dynamic society than terms of traditional or customary use that were no longer shared or were dependent on an authority that was losing its legitimacy. Quantification had another far-reaching consequence—it made much of Western thought more abstract and self-critical.[43]

Abstraction

The representation of reality through numbers is not intrinsically more abstract than representations based on words or pictures. But when numerical representation pervades public life and becomes subsumable within an entire calculus of quantitative relations, such as geometry or algebra, it reinforces representation that is precise and conventional, yet cold and remote from the contradictions of ordinary experience. Such clarity and remoteness are characteristic of scientific conceptions and promote their public, objective character. The meanings of scientific concepts are consciously created and arrived at by the consensus of the scientific community. This consciousness encourages science and related realms to become self-critical. It makes scientists aware that they are creating symbols or models whose meanings are negotiable and that they are not creating reality or affecting the reality represented (as is believed to be the case in ritual and magic). The consciousness of the role of symbolization, assisted by quantification, increases intellectual experimentation in other modes of thought.

Human observation and experience as the measure of knowledge, the quantification of facts and values, and the emergence of science were not direct causes of one another, nor did they immediately dominate Western thought. They were consonant with ideas that interpenetrated and reinforced each other as modernity developed. They were also related to the changes in the conception and uses of space, time, and territoriality. The development of a more metrical and quantitative conception of space, place, and time facilitated movement, coordination, and control of activities over the earth, because it thinned out the everyday meaning of space and time. Space and time have become abstract frameworks to which events and experiences are only contingently related.

Time

The changes triggered by modernity have been rationalized through the belief in progress. People expect that change is for the better. Although older societies held that the future might be better, it was a belief rooted in cycles of good times and bad, a belief that what has been will occur again, a belief that if something entirely new occurred, it was only through divine intervention. The Western notion of progress, however, means belief in a continuous secular change to new, better, and heretofore unavailable and even unforeseen conditions.

The idea of progress as well as more mundane experiences of change in modern society rely on an abstract metrical notion of time. The abstraction of time is supported by the development of accurate and portable timepieces. This does not mean that time is independent of events, for it still must be measured in terms of changes in the material world, such as the path of the sun or the movement of a hand on a clock, but it does mean that the measure can have unrestricted referents because it can serve as a neutral, public, and objective measure to mark duration and sequence of events. In fact, instead of events marking time, metrical units of time have come to define events, as when in factory life the workday begins at 8 A.M. and ends at 5 P.M. This conceptual separation of time from events allows it to have value in itself, which is extremely useful to capitalism, as many kinds of work are reckoned in terms of buying and selling time.[44]

Space

As public, metrical, time became freed from experience and context, so too did geographical space. Few premodern societies, including medieval Europe, made systematic use of a spatial grid to represent the world. The Chinese possessed a coordinate system, but theirs seems to have been infused with spiritual meaning.[45] Greek geometry and cartography were more concerned with solid objects than with abstract space.[46] Yet some Greek mathematics and cartography had the potential to address earth and celestial space as one metrical system. Ptolemy's discussion of projections and coordinates comes as close as any premodern formulation to an abstraction and metrication of space. In this regard, it is significant that Ptolemy's system was available to Christian Europe by A.D. 1150 but was not used by medieval geographers. Rather, medieval scientists applied it to the heavens, and it became infused with astrology. For earth space, medieval geography often substituted a mystical view derived from church doctrine and portrayed in the well-known T-O maps.

Not until the fifteenth century, with the awakening of interest in navigation and trade, was Ptolemy's system of mapping the earth rediscovered: 1405 saw the first translation in the West of Ptolemy's *Geography*. Soon thereafter, cartographers represented space in terms of coordinates such as longitude and latitude. The use of a coordinate system to represent space depends on imagining the globe not as an amorphous topography but as a homogeneous surface "ruled" by a uniform grid.[47] Abstract metrical space was joined to abstract metrical time to form a general, yet precise, public frame of reference in which human experiences were contingently located.

The mapping of space in terms of coordinates was only one instrument expressing space as an abstract framework for events. The other was perspective painting. Before the fifteenth century, the concept of space in painting was overwhelmed by the concepts of position and size. The size of an object and its position in the panting indicated something of its importance but nothing of its actual location in a geographical reality. In perspective painting, the events depicted were literally painted on a pre-existing coordinate system representing space itself. Perspective painting and Renaissance cartography reinforced each other. Artists were aware of new cartographic methods—and cartographers were often artists. Their connections may have been so close that Ptolemaic rules for map projections in the Almagest may have been adopted by Alberti, one of the founders of perspective painting, in his construction of perspectives.[48]

Territoriality

Developments in cartography and painting were of great importance in conceptualizing and rendering a public abstract spatial view from nowhere. But this view was undergirded by spatial practices using territoriality at all levels to move, mold, and control human spatial organization. These activities make territories appear conceptually and even actually emptiable and fillable, and this presents space as both a real and emptiable surface or stage on which events occur.

Using territory as a mold or container for social organizations occurred since the beginning of recorded time, but it increased in intensity and scale with the discovery of the New World. The New World, and especially North America, presented European powers with a vast, distant, unknown, and novel area. This meant that, even with the limited technology and political power at their disposal, Europeans could still "clear" space and form territories at all geographical levels, with an intensity that was impossible to match in the Old World. Evidence of European thinking of territory as emptiable space is indicated by several events:

- North American charters and grants, which delimited European claims by using the abstract metrical lines of latitude and by their provision for a hierarchy of administrative subterritories long before the land was surveyed and settled.
- The European claim that the land was virtually uninhabited and European displacement of most of the New World's aboriginal population.
- The change of definition of community as one in which admission required only residence within the community's territory.
- The shift from a form of representation in which a community was thought to be an organic entity with a common interest (and thus needing only a single representative to give voice to its needs) to proportional representation based on periodic censuses, in which the community is thought of more as a collection of individuals than as a unified body.
- The United States Constitution, in its use of areal representation as a device to divide factions and balance power.
- The territorial partitioning of the Western lands.

The treatment of territory as an emptiable and fillable container is even greater in the present day. The same treatment is also evident on a smaller scale, such as in neighborhoods and buildings. The primary difference is that here (unlike on the political level, which because of coordinate maps could be presented as an empty space to partition and then fill), the small-scale and architectural levels were seen as emptiable only after they were thinned out—after each thing was put into a separate place. This thinning out did not mean a lowering of density. On the contrary, it meant first isolating and segmenting the activities to be contained, and this often meant multiplication and intensification of territory.

These small-scale transformations are revealed at one level in the clearing of city streets. European medieval and Renaissance town streets (and those of other premodern societies) were filled with the hustle and bustle of numerous activities. Merchants peddled wares, beggars panhandled, families socialized, town criers spread news, and public trials and hangings occurred. But as commercial interests became more important, access to public spaces like roads and squares was restricted. Rules prohibited merchants from hawking their wares on the streets, restricted beggars to certain locations, prohibited social gatherings, and in general limited the use of streets and roads only to transporting people and goods from one place to another. As the activities in streets became thinned out and cleared for transportation, the city became economically differentiated; waterfronts were cleared for warehousing, stock exchanges were established in accessible areas; and with industrial capitalism, the city took on its modern form of separate residential areas and manufacturing and central business districts.[49]

A corresponding thinning out of place through territorial control also happened in domestic architecture. Before the commercial revolution, the major architectural organization of even the largest house was not based on assignment of specific functions to specific rooms or places. Rather, homes were subdivided into rooms that could serve multiple functions, and in most cases each room led into another, for few central corridors or hallways existed solely for access.[50] But just as activities in streets and shops thinned out, so they did within the house, which soon contained central corridors for movement and access to rooms.[51] The thinning out and eventual conceptual emptying of places occurred most dramatically in work environments. Workers now had to labor away from home in a place and on machines they did not own. This meant that capitalists and managers had to define, supervise, and control minute details of the work process. Workers came to the factory at specified times, and they worked at specified places on specified machines for specified lengths of time. They could not leave their work stations without permission, and they could not vary the place of their work. Workers were constantly under the gaze of supervisors, and workers' bodies were little more than appendages to machines.

Since work was now separated from home, and members of the family were now physically separated during the day, other geographically distinct institutions arose to supply old and new services for households. New territorial forms sprang up for production, consumption, and surveillance: school buildings contained classrooms and desks with assigned and ordered seating; prisons were subdivided into cells and cell blocks; hospitals and asylums contained wings and ordered rows for rooms and beds. Common to many of these structures was an arrangement of places that facilitated hierarchical access, supervision, and control.

The increasing division of labor and spatial segmentation of elements of life thinned out activities in space. Only a few minutely defined events were to occur in any one place. In many cases, such events were so unusual that the structures designed to house them were not suitable for anything else. When the events were no longer housed in the building, the building remained empty, abandoned, unless it was drastically renovated. But most events, although minutely subdivided, segmented, and thinned out, could be housed within a general type of architectural shell if the interior partitions were flexible enough to allow rearrangement. The same basic structure, with only slight modifications, could then serve a variety of purposes. This was the case in the panoptical designs for institutions, such as hospitals, asylums, schools, prisons, and factories. Even apartment buildings were designed with movable interior partitions so that occupants could compartmentalize their living space as desired. This sense of a flexible and conceptually emptiable container has been incorporated into the language of architecture. Architects still build buildings, but they are now called *volumes* or *spaces*.

The power to create versatile architectural forms and to minutely subdivide, organize, and reorganize every aspect within them, made the built environment both a conceptually and actually emptiable and reusable space or container. Seeing and using space as a container at the architectural level merges with the awareness of political geographical space as a surface or volume in which events occur. The same sense pertains at both scales, because modern society possesses the power, through territorial control, to repeatedly empty, fill, and rearrange events and meanings in space. It means that events and space are conceptually separable; they are only contingently related. People, things, processes, and meaning are not anchored to a place—are not essentially and necessarily of a place. This conceptual separation of space and things at the practical level imparts an abstract quality to space, which in turn reinforces the role of a geometrical space as the public view from nowhere from a dynamic and complex society. Territorial partitioning, with private property (or "alienable parcels") as a special case, gives meaning to David Harvey's comment that "one of the ways in which the homogeneity of space can be achieved is through its total 'pulverization.'"[52]

Quantification and territoriality are important mechanisms helping to propel us to an objective, public, and geometrical view from nowhere. But these factors and others also work in the opposite direction, to provide a view from somewhere, so that we can differentiate self from world. The dynamic modern world and its public objective space make the personal sense of being in place difficult to share. We need places of our own in which to build our own worlds. In Yi-Fu Tuan's terms, as the external world becomes more segmented, so too does the self. Territoriality provides the possibility of anchoring the personal sense of being in place to fixed locations in space. But because this personal sense of being in place is not shared—and because territorial partitioning may even reinforce its isolation—it becomes private, subjective, and even idiosyncratic.

•

The views from somewhere and nowhere establish one important set of epistemological or methodological tensions in the modern world. An equally important though ontological set arises over the debates about the primary forces that affect our actions and that help constitute our places. Are our actions (at particular times and places) driven primarily by forces from the realm of nature or from the realm of social relations, or from the realm of meaning? Or are none of these determinate because of our free will? The question of what forces drive us is as important as the issue of how to view them. In our everyday lives, we find that places of consumption integrate and alter elements from forces and perspectives. But before I examine the structure of these places, I will expand the framework by considering how one particular realm of force—agency—is especially sensitive to the shift in perspectives from somewhere to nowhere.

· 3 ·
The Problem of Agency

Forces assist the development of perspectives from somewhere to nowhere. But perspectives also exert an influence, for they affect the appearance of what can be seen. The personal sense of place appears most clearly from a perspective from somewhere along the discursive and scientific axis (this somewhere rises above the preconscious to include awareness). The perspective from virtually nowhere along this axis presents place primarily as a location in a spatial system that can be described geometrically. In addition, other perspectives draw into focus yet other facets of space and place; they do the same to the realms of force. This is an important point, for one of the major realms—that of the free agent—seems to all but disappear as we move from somewhere to nowhere; it is replaced by a structurationist sense of agency. (This is why agency is not a separate realm of force in figure 1.) Exploring the difficulties that all of the perspectives (except from somewhere) have in focusing on free agency helps elaborate the relational framework and helps us understand the significance of place to morality and responsibility. Place can provide a perspective from somewhere, which is the position that brings free will into sharpest focus—and belief in free will (or the ability to do otherwise) is essential to moral choice and responsibility.

Free agency, or free will, is part of theological concern with the nature of creation, of good and evil, and hence of human responsibility. In a secular society with a strong social science tradition of viewing things from virtually nowhere, these questions are difficult to address directly, and so the problem of free will becomes reduced to the narrower question of being free to choose or to act differently. This is the sense of free agency addressed in this chapter and the next (under the discussion of reflexivity), but the more general sense of free agency appears again in my discussions of morality in chapter 8. The analysis of the relation between somewhere, nowhere, and free agency begins with a consideration of the kinds of facts that can be seen more clearly from one or another perspective. This discussion of agency is brief, but I have made it into a separate chapter because the topic deserves special attention and because the discussion serves as a natural transition from perspectives to forces.

Autogenic Actions and Impact

When we discuss how a perspective affects what is viewed, it is difficult not to make the claim that there really is nothing out there to view; all we possess are perspectives. This position is, of course, a form of idealism, and the opposite assertion—that there are indeed things to view that exist independently of perspective—is a form of realism. I am a realist in that I believe that things really are out there to know and that these are independent of our thoughts. Yet we cannot know them without thinking about them, and thus in a Kantian sense their appearance will always be affected by the lens through which we see them. Neither realism nor idealism can be justified in any absolute sense: they are simply articles of faith. Still, it is important to be more specific about how a belief in the reality of forces can be formulated in such a way as to allow forces and perspectives to be mutually interactive. A commitment to this kind of connection is important to mention at the outset, because what I say about free agency can be misunderstood to mean that it does not exist but is, rather, a figment of the perspective from somewhere. My position is that it does exist; it must exist for humans to be morally responsible. But it can be seen, and then only fleetingly, from the perspective of somewhere.

Thus far, the qualities of perspectives I have focused on are their degrees of subjectivity and objectivity—their view from somewhere and nowhere. The distinction between subjectivity and objectivity draws attention not only to the awareness molded by circumstances but also reflects a real difference among the forces we experience. In other words, the epistemological issue of subjectivity and objectivity corresponds to an ontological issue, which is another way of saying that epistemology and ontology are interdependent. Subjectivity and objectivity apply not only to how we see things but to what we see: forces from each of these realms might be more subjective or more objective by their very nature. The more subjective forces could be those that seem to have their origins within ourselves.

These autogenic forces, as Susanne Langer calls them, seem to be initiated by the activities of our nervous systems and include our deepest feelings and emotions. The more objective experiences are those that seem to originate outside ourselves, and we sense this external quality because they affect us as impact.[1] These two can of course be interconnected, in that events from the outside can trigger autogenic feelings, and these feelings can affect how the impacts are experienced. This means that the three realms of forces are each divided into an autogenic part and an impact part. The two parts are related to perspectives in that the more objective perspectives, such as the sciences, focus more clearly on the more objective facts: those that seem to be outside ourselves and that are felt as impact. Subjective perspectives are related to subjective facts of experience or to the autogenic part of forces.

The distinction between objective and subjective is found in varying degrees in each of the realms. That is, many forces from the realms of meaning can appear to be outside ourselves and experienced as impact, and many others can be experienced as autogenic. The same can be said for social relations and nature. One kind of fact may occur more often in one realm than in another, but at this stage it is impossible to be more definitive about the matter, except to say that the realm of free agency, as we shall see, may be most clearly visible from the perspective from somewhere, because it is composed primarily of autogenic forces.

This distinction between subjective and objective forces provides a foundation to the claim that the distinctions I am making are not entirely matters of perspective but also matters of reality.[2] The implication of a division of experience into subjective and objective is that those things we experience most intensely as impact—as coming from outside of us—come into focus as we move toward the perspective from nowhere. But viewing from this perspective means that we see our entire selves only this way—as something outside ourselves, as impact. From this view, we have difficulty observing our own, autogenic, sensations. It is as though feelings and emotions are seen secondhand.

Free Agency and Objectivity

One of the clearest illustrations of how we have difficulty seeing the subjective from an objective perspective is the way our sense of being free agents (or being free to choose, which is implied in the structurationist terms "agent" and "structures that enable") diminishes as our perspective moves from somewhere to nowhere. I will illustrate this with an example that focuses on the connection between freedom and choice. The example, borrowed from Thomas Nagel, is typical of choices we encounter in our daily lives.

Suppose I have just completed my dinner in a restaurant and I am deciding whether to order fruit or ice cream for dessert. I realize that the fruit would be better for me, but instead I decide to order ice cream. As soon as the ice cream arrives, I berate myself and say that I should have ordered the fruit. In this example, I really believe that I had a choice. I really believe I could have ordered the fruit instead of the ice cream. Or I could have had no dessert at all. In this sense, I sincerely believe that my choice was free—that it was I who decided and that my choice was not caused "in advance."[3]

Choice is free only within the constraints imposed by structure. In my case, various structures, including the menu, the restaurant, and even capitalism, presented an array of alternatives, of which only two interested me—ice cream and fruit. The action of ordering dessert helped reproduce these structures, including the menu, the restaurant, and even capitalism. I was not free in the sense that I could order something not on the menu (though asking for it might eventually have changed the menu and other structures). Nor is this selection of choices particularly important to the conduct of my life. Indeed, it can be argued that certain structures, such as capitalism, present the illusion of choice to disguise the fact that these structures constrain behavior. It might be argued, then, that we should make choices that change structures so that we can have real alternatives and, thus, real freedom.

These issues, though politically important, do not address the logical problems of choice. Freedom can never mean the ability to choose without limit. My interest here is in the interpretation of what it means to be able to choose, given constraints (and as we shall see, transforming structures does not introduce anything new to the question of choice). This question of free choice is important in its own right, because it is the basis of moral responsibility. I intentionally selected a simple and ethically neutral example (which most social theories would not deign to consider) precisely because, if choice is problematical in such a case, it would be even more so for really important decisions.

Returning then to my example, I can ask, Does it mean that my choice between ice cream and fruit is in some sense free, in that I could have made a different choice? The difficulty here is that, as soon as I try to examine why I chose ice cream and not fruit, I begin to distance myself from my decision and to seek reasons—which eventually become causes or forces—for my action (a position that Anthony Giddens also takes).[4] I could say that my craving for sweets overpowered me or that I like to flirt with danger. And if you instead of I were to explain what I did, you might see my choice as having been controlled by other factors, such as the power of advertising over that of common sense; but you too would be seeing me from the outside, as an agent driven by forces beyond my control.

The awareness of my action and the attempt to understand it and see it from outside myself establish a tension between my belief that I caused my action to happen and a view that sees my action as caused by other forces. The tension can be put more generally. If I chose *x* over *y*, and if I believed I really had a choice, that means I could have selected *y* over *x*. Moreover, this means that my choice was not determined in advance by my desires, my personality, my beliefs, or anything else that could act as a force (either internal or external) compelling me to select *x*.[5] But if my selection of *x* over *y* was not uncaused—and how could it be uncaused if I believe it was I who caused it?—then what are we left with as an explanation of my choice? How was it caused? It seems that no matter how strongly I believe that I had control over what I did, there really are only two ways of explaining it. Either I must accept that my choice was caused by such things as my desires, personality, beliefs, social forces, and so forth and that my feelings about free choice are only an illusion (though an important one that itself is worthy of explaining), or I must accept that my choice was determined solely by me, which is the same as saying it was caused by "my doing it."[6] But this alternative—my doing it—is not really an explanation at all. It simply says that I undertook an action. And if I leave it at that, then I must accept that my action—my doing it—was uncaused.

What happens, though, if we transform the structure so that it provides other and better choices? Does this mean that we now have shaken ourselves free of other causes and really exercise choice? The answer is that altering structures does not change the problem. From our own internal perspectives, we may feel that we made a free choice to change a social structure, but from outside ourselves, we would still seek to explain why we undertook such a transformation. And again we would be left with the alternative that we simply did it (and thus were the causes and were ourselves uncaused) or with the alternative that our actions were themselves driven by other, larger and perhaps newer, forces or structures, which now replace the older ones.

Free Agency and the Laws and Rules of Social Science

If the sense of free agency becomes blurred when we step back and explain our actions in everyday life, it certainly loses its clarity when seen from the perspective of science and of most of the social sciences. The latter, although it never really addresses these issues head-on, assumes that our behavior is caused, because scientists examine behavior from an objective perspective and they are in the business of finding causes and proposing explanations. This search for causal relations is reflected in the meaning that social science gives to the term *agent*—a meaning that coincides with a commonsensical definition. An agent is someone who is the instrument of a particular force, as an agent of the FBI or an agent of capitalism. As an agent of the FBI, our actions are constrained by our office, by U.S. laws, by customs and norms, and perhaps even by certain "laws" of human action.

Candidates for such laws from the realm of social relations are Max Weber's generalizations about bureaucracy and Roberto Michels's "iron law of oligarchy."[7] These laws state that, once we are part of a bureaucracy, we are driven to behave in ways that perpetuate that bureaucracy. These laws, although instantiated by the actions of individuals, appear to be beyond the control of any individual and are basic to a bureaucracy's survival. We could also look to the realm of meaning (and nature, for that matter) to find forces that make us their agents. Mental forces common to all humans, such as those proposed by Freud, could constrain our behavior in ways we cannot control.

Laws of human action, if they exist, would disclose forces that we cannot alter or avoid. It is possible to violate or change rules and norms, yet these too can be overwhelming and compelling forces. Most social scientists do not discuss which kind of force they have in mind. The properties that both laws and rules disclose have also been called *structures*, but the social science literature seems to reserve that term for forces that emanate from large-scale social institutions. Much of the work of social science is to discover such structural forces, and from this general perspective there is little room for a free human agent.

How does social science address the situation of an agent who is under the influence of forces of varying strengths, and whose behavior is caused—but not (evidently) determined by any single one?[8] The FBI agent, though embedded in a role, may also have other roles. Perhaps he or she is also a scholar, a parent, a consumer. Any given situation, then, could be affected by more than one role or force; and it may be unclear which one is primary. Social scientists could describe such a situation in several ways. They would note that all theories, laws, and models are about potential relations, which are conditional and apply only to the domain they specify. Thus not all of the actions of an agent would be explained by his or her position in a single institution. Other forces could override these.

These same qualifications apply to natural science generalizations. Our knowledge of combustion allows us to say that, if everything else is equal, striking a dry match with sufficient force against an abrasive element will make the match burst into flames; or our knowledge of changes in state allows us to say that a body of water will turn to ice if its temperature drops below zero degrees Celsius. These theories about combustion and freezing identify potentialities that could occur if the systems are closed. We all know, however, that a dry match struck with sufficient force will not necessarily ignite, nor will water necessarily freeze at zero centigrade if unforeseen factors intervene—a blast of wind or a lack of oxygen in the case of the match, turbulence or impurity in the case of water. The same applies to the FBI agent, whose role as an agent may be superseded by his or her role as a parent, friend, or political liberal.

Just as physical science theories cannot predict whether any one of a virtually infinite number of other systems may intervene to affect combustion or freezing, so social theories cannot predict whether other systems beyond its domain will interfere with its subject matter. Even so, social and natural scientists maintain that such activities are caused, though not necessarily by the forces outlined in the theory. These unstipulated forces are contingent, insofar as they are not significant if the system is closed. (Many good theoretical reasons argue for layers of necessary and contingent relations.) These open and contingent relations then become probabilistic, and the events or human actions are viewed as caused but not determined.[9]

Many social science theories even begin with the assumption of contingency and then outline the constraints within which the contingency operates. Modern economic theory is a case in point: economists argue that we purchase products to satisfy our needs, and they do not say how these needs are established. The formation of these needs is external to economics and is contingent, though perhaps caused by psychological factors. Rather, economics is concerned with demonstrating how certain types of structures, such as the free-market system, constrain the exercise of our self-interest. Even if we wish to purchase *x*, because it will satisfy a need or desire of ours, the price of *x* will be determined by forces beyond our control: by the intersection of the supply and demand curves generated by other consumers and by producers. And when economics discusses whether we will in fact purchase *x* at a particular price, it does so in terms of other constraints that we face: the size of our budget, the availability and price of *x*, and our preference for other items.[10] The market system, then, is imagined to be an efficient means for allowing us to satisfy our needs or self-interests, but why these needs exist in the first place is not part of economics. And even though economics describes this as a "free" market system, which allows us the "freedom" to satisfy our needs, the needs themselves, as the word implies, are hardly uncaused, and thus we are driven by forces that are simply not addressed by the market system.

Even in economics, then, we are not free agents—whose behavior is uncaused. Rather, we are free only to satisfy our needs, which act as forces directing our behavior. This is about as close to freedom as the perspective from the social science position of nowhere permits. Even the structurational formulation of social theory does not address free agency (though it may imply that it does by embedding its concerns within Marx's general phrase: we create history but not under conditions of our own choosing).[11] Within this framework, our actions are still caused (though obviously not simply by social structures).

Because of its emphasis on rules of conduct and unintended consequences of actions, structuration theory adds several important insights: forces or structures constrain and enable human actions (as the English language, with its vocabulary and grammatical rules, enables us to communicate, yet constrains what we say); these forces or structures must be instantiated by human practice; and such practices can alter the structures (so that the power of English exists only because it is used, and in using it we ever so slightly alter it and contribute to its structure). However, the desire to speak English and the content of speech are still not results of free agency. *Enabling* (in the constraint enabling dualism of structuration terminology) is not synonymous with an *agent* whose actions are uncaused by external factors. Rather, *enabling* means that specific actions are not completely caused by particular forces in particular social structures. Still, this same behavior could be caused by the conjunction of other forces from other kinds of structures.

Most of the social science perspective (including economics), since it is a view from virtually nowhere, cannot accept the existence of free agency, because free agency must be uncaused, or caused by our will, which amounts to the same thing. Rather, social science interprets our internal sense (not anchored literally in a preconscious somewhere but requiring some awareness) that free agency exists as a reaction, an epiphenomenon, which itself needs to be explained, perhaps by describing it as a sensation that accompanies the confluence of competing forces. This doubt about free agency that arises along the discursive and scientific axis also arises in the move from subjectivity to objectivity in general, even if it involves only ourselves questioning our own choices and thus seeing ourselves from somewhere else. The difficulty in observing free agency is an example of how subjective facts (those that feel autogenic) become increasingly difficult to view as we move to a more objective position. According to Thomas Nagel,

> Something peculiar happens when we view action from an objective or external standpoint. Some of its most important features seem to vanish under the objective gaze. Actions seem no longer assignable to individual agents as sources, but become instead components of the flux of events in the world of which the agent is a part. The easiest way to produce this effect is to think

> of the possibility that all actions are causally determined, but it is not the only way. The essential source of the problem is a view of persons and their actions as part of the order of nature, causally determined. That conception, if pressed, leads to the feeling that we are not agents at all, that we are helpless and not responsible for what we do. Against this judgment the inner view of the agent rebels. The question is whether it can stand up to the debilitating effects of [this external] view.[12]

The appearance and then virtual disappearance of free will (and I assume it exists, even though it is difficult to see) illustrates how forces can appear very different when viewed from one or another perspective. But the movement from somewhere to nowhere is not simply the result of mental effort. It requires the assistance of forces from the other realms and the manipulation of geographical place and space. The capacity of the framework to incorporate these complex interconnections is another reason it is relational. I turn next to the other realms of force, which, though affected by perspectives, are not in danger of vanishing when our gaze shifts from somewhere to nowhere.

· 4 ·

Forces from the Realms of Meaning, Nature, and Social Relations

Somewhere to nowhere encompasses a wide range of perspectives. But what really exists out there for us to view? What are the major forces affecting us? The modern intellectual world poses four realms of force: *nature,* or the forces and relations that the natural scientists identify; *meaning,* the forces of the mind; *social relations,* which include race, class, gender, and bureaucracy; and the force of *free agency.* (As is clear from chapter 3, the realm of free agency is the most difficult to keep in focus and hence does not appear as a separate realm in figure 1 but, rather, penetrates meaning and social relations.) Although the forces are real, their divisions into realms are partly a conceptual construct that Western culture has developed to divide the intellectual labor, so to speak. These divisions are not found in every society, and they have only emerged over time in ours; and even in our own society, it is assumed, overall, that there is a unity to the world and that each realm interpenetrates and helps constitute the others.[1]

Yet specialization tends to make most researchers examine forces primarily from within only one of the realms, to the point where forces from other realms are seen as something like background material or else are completely disregarded. Thus most of natural science does not consider human nature; most of sociology, economics, and political science do not consider nature, and give only perfunctory attention to theories of the mind; and those interested in psychology and intellectual history do not emphasize social relations or the natural world. We should not jump to the conclusion that all theories within a realm look alike. Even though sets of theories may draw on the same forces, they can work these differently and thus come to very different conclusions.

A few theories attempt to construct bridges between one realm and another, as in social psychology's efforts to connect the social realm and the realm of meaning, or in ecology's linking of natural and social forces. And a very few theories—such as Marx's or Freud's, which have been called totalizing—attempt to bridge all three realms, but even these use one realm as a base from which to examine the others. Indeed, the most protracted interpretative issues in social science concern the degree to which a totalizing theory draws on other realms. For example, Marxist theory is most concerned with social relations, but can it also be considered a naturalist theory, deriving its primary force from the natural world? By the same token, Freudian theory is most concerned with the realm of meaning, specifically with the construction of personality, but does it also find its fundamental force in the natural realm through the guise of biological drives and instincts? The fact that these questions can be posed points to the prominence of these realms as the intellectual foundations of contemporary theory.

It would be imprudent to make too strong a connection between these realms and conventional academic divisions. Although the realm of nature has come to be defined by the domain of the natural sciences, the realms of social relations and meaning do not correspond neatly to the boundaries between the social sciences and the humanities. For instance, psychology, cultural anthropology, and linguistics, which many would call social sciences, and even intellectual history, which some would regard as a social science, are principle sources for theories about the realm of meaning.

Diverse answers, then, are expected to our question about the fundamental forces that affect our lives, but at the core, these answers draw upon one or the other of these realms. These realms, and the partial and conflicting theories that they generate, have fragmented the intellectual world and placed it in something of a deadlock. At least this is the way it appears if one is not an advocate of only one of these totalizing theories.

A connectedness and even a hierarchy among the forces can be imposed at a superficial level by assuming a metatheoretical position, such as that found in the idea of reduction. This idea is an undercurrent to many parts of the natural and social sciences that advocate naturalism.[2] Reduction in the technical sense means that human behavior, which includes the realms of meaning and social relations, is built upon natural processes and thus ultimately is part of and controlled by natural forces. Furthermore, this means that, if there were generalizations or laws about human nature, these would themselves be explainable by the laws or generalizations of the natural sciences. Reduction also suggests that the realm of meaning, and its psychological forces, is more fundamental than the realm of social relations, which concerns itself with group or aggregate behavior.

But it can be argued that social theories are as essential as natural science theory to an explanation of human behavior. Human nature involves novelties (or entelechies) that are not reducible to the facts of natural science. Although we cannot defy the laws of nature, what we know of these laws comes from our conceptions of them and thus is partly molded by our minds and our social organization.

Such intellectual tugs-of-war have led to a deadlock among theories and brings us back to the issue of perspectives and their relation to forces. The following sections give an overview of the relations among the realms of meaning, nature, and social relations. Instead of focusing on the particular theories within the realms, which is too vast a problem for a single volume, I try to provide a sense of the intellectual deadlock by examining those theories that have something to say about the relation of their own realm to other realms. The purpose here is not a review of contemporary theories but an examination of how contemporary theoretical discourse assumes the existence of these realms of forces. The theories examined by their very nature employ degrees of objectivity; thus free agency is not part of the discussion, since it cannot be seen from an objective perspective. Once *free* is eliminated from *free agency,* we are left simply with agents who are instruments of forces and thus pervade all of the realms—although, as Anthony Giddens reminds us, agents still help to constitute such forces.[3]

The second section reconsiders the connection between forces and perspectives. Forces are known through theories, and theories lead to perspectives, and until this issue is discussed directly, we will not challenge the validity of any of these theories and will assume that what they state about the existence of forces is plausible. The last section covers space and place as forces pervading each of the realms: with the exception of the natural sciences, most theories do not address space and place explicitly. So instead of explicating their role in these theories, which would be tantamount to making them more geographical, I discuss the relation of space and place to these realms separately at the end of the chapter. Let us turn then to the overview of each realm.

Connections among the Realms

The Realm of Meaning

Many examples exist of our minds providing the primary power for our actions and for our transformation of the world. Some of these theories are simple, others complex. Most of them consider the connection of the mind to social relations but virtually ignore its connection to nature (Max Weber's work being a prime example). Here, I examine three theories that offer different conceptions of the mind and that consider the mind's effect on nature and social relations. The first is the position expressed by Lynn White, which is the simplest of the three theories and represents the way many scholars portray the power of the mind—as values or ideas that propel and enable us to act and to transform the world.[4] The second theory, from Claude Lévi-Strauss, explores how the mind constructs a view of nature and social relations, albeit a static one. The third, from Sigmund Freud, examines how the forces of the mind are organized and continuously draw upon elements from the biological portion of the natural realm to develop complex systems of social relations. Because Freud's theories are so wide-ranging, they serve as a basis for theories of the mind and offer a transition to our discussion of nature.

White's position is that the responsibility for our ecological crises can be laid at the doorstep of Christianity. Christianity teaches that humans have dominion over nature. This teaching has come to be so much a part of Western values that people act without concern for nature, which they simply treat as a resource to be dominated and transformed. "Christianity . . . not only established a dualism of man and nature but also insisted that it is God's will that man exploit nature for his proper ends." White further claims that the "victory of Christianity over paganism was the greatest psychic revolution in the history of our culture." This victory instilled the twin ideas of domination over nature and progress, which see the cumulative weight of these actions as improvements. It is true that Christianity, as an idea by itself, did not do the transforming. Rather, it acted as a force that set in motion a complex web of social practices, including the development of science and technology, through which this mental commitment operates. Still, the motor for the transformation is squarely within the realm of ideas.

White's commitment to the mind as the primary source of change is also revealed by the fact that he believes the primary means by which we can mend our ways is to change our values: "We shall continue to have a worsening ecologic crisis until we reject the Christian axiom that nature has no reason for existence save to serve man." White proposes as an alternative the view of Saint Francis of Assisi, who "tried to substitute the idea of the equality of all creatures, including men, for the idea of man's limitless rule of creation." But, while White believes humans can change, this does not necessarily mean that we are free agents. After all, we may choose an alternative for reasons that are compelling, and these reasons would then act as causes or forces.[5]

White's position can be attacked on several grounds, including the particular issue of whether this interpretation of Christianity is the one held by most in the West and the more general issues of whether any set of values acts independently of forces from other realms and could be strong enough to provide the primary impetus for vast environmental change. An excellent example of a challenge to White based on these more general issues is Yi-Fu Tuan's comparison of Chinese Taoist attitudes toward nature and China's effects on the land, on the one hand, and Western Christian attitudes and uses of the land, on the other.[6] Tuan's assessment of the Western view toward nature does not differ from White's. They both hold that the West expects that humans should dominate nature, whereas the predominant view in China—the Taoist view—is one of harmony with nature. Yet in comparing the scale of environmental changes caused by both civilizations until the rise of the industrial revolution, Tuan is hard pressed to find a difference: the transformations wrought in the two traditions are comparable.

Even if we accept the argument that Christian values have influenced the transformation of nature, other complicating factors abound in interpreting the specific form of these particular values. For example, Protestantism, and especially the Puritan ethic, has encouraged individualism and acquisitiveness, which are necessary values in the modern consumer world. Consumers must believe not only that they should transform nature to meet their needs but that their needs are unlimited and are worth satisfying at all costs. Is this unbridled acquisitiveness a product of particular forms of Christianity, or is it a product of other forces? Another means of attacking positions like White's is to argue that Christian values in particular—and the realm of meaning in general—are not autonomous but are dependent on other realms. Particular values are outgrowths of particular social relations, and it could be the latter that are the primary factors transforming nature.

Lévi-Strauss's theory of structuralism addresses the effects of the unconscious structure of the human mind on nature and social relations.[7] He argues that the mind is programmed to work by forming categories of extreme oppositions and mediation, which are then repeated at different levels through analogy. These operations are most clearly revealed in the mythical structures of preliterate societies. A myth's content, in a sense, is secondary to its structure, which reveals the inner workings of the mind. For the most part, these mental processes are unconscious. According to G. S. Kirk, Lévi-Strauss does not claim to show how people think in their myths but "how myths think themselves in men, and without their awareness."[8] Oppositions reflect irreconcilable dualities, such as life/death, human/god, good/evil, male/female, up/down, and front/back. Such oppositions provide a scaffolding, or structure, to organize ideas about nature and society.[9] Within this framework, social activity becomes interpreted as acts of communication generated by this structure.

Lévi-Strauss's examples are often extremely complex and assume that readers have a considerable background in the ethnography and environment of a people. Therefore, I will apply his method to a Western view about people and nature, with the understanding that the method seems to be best exemplified by simpler, preliterate societies. A prime illustration is provided once again by Christianity. To a Lévi-Straussian, many of the characteristics of Christ—his being the son of God, his having been immaculately conceived, and his resurrection—are a product of our attempts to reconcile dualities. God is immortal, unchanging, and perfect. Humans are mortal, changeable, and imperfect. These extremes can be reduced by developing a mediating concept that reduces the tensions between these and related categories. A primary mediating concept is Christ the man/god who dies and yet is immortal. The structure can develop further through other intermediaries, such as priests, bishops, archbishops, and the pope. Other intermediate positions could be developed to close the geographical gap between heaven and earth. Earthly holy places and sacred shrines can bring us closer to heaven. After death, we might need to go through intermediate steps before attaining heaven.

Another case, closer to home, is an interpretation by Tuan of one of our contemporary conceptions of wilderness.[10] Wilderness can be thought of as part of a modern Western opposition. Recently, wilderness has come to be valued positively. It is natural, pure, and unchanging. Before this, it was thought of as hostile and something to conquer. It is also a place with little or no human activity. The relative absence of humans is what makes wilderness pure. Its opposite is areas of greatest human habitation and control, or cities. These are almost completely human-made environments and are often seen as unnatural, impure, and impermanent. Still, these opposing areas contain elements of each other. Wilderness requires human intervention: it must be protected by humans, and it is there for humans to visit. And cities are not completely immune from nature: natural elements can still wreak havoc. In order to lessen the opposition, we create intermediate categories and places, such as suburbs, city parks, and zoos.[11] This analysis can continue almost indefinitely, but this may be sufficient to illustrate the position that the power of the mind creates oppositions and mediations that order the natural and social worlds. Because the structure involves oppositions, our reactions to nature and culture are marked by deep ambivalence.

White and Lévi-Strauss both believe that the realm of meaning is the primary locus of power. White emphasizes the power of one idea. Lévi-Strauss presents a mental structure that cannot be escaped. Lévi-Strauss's model is primarily static, whereas White's is dynamic.[12] As the founder of psychoanalysis and other branches of psychology, Freud focuses his attention on the nature of psychological forces. Although he saw the principal sources of psychic energy as stemming from our biological drives and instincts, his efforts were directed toward understanding the way in which the mind draws upon and transforms these forces and how they lead to social relations. In his most succinct discussion of these relationships, *Civilization and Its Discontents,* Freud argues that civilization creates frustrations by placing obstacles in the path of the immediate gratification of our desires and drives.[13] This might appear to mean that social relations dominate psychological and even biological relations, yet civilization itself is a product of displaced psychological forces, so that the effects of civilization are something like feedback loops within a psychological system.

In its simplest form, Freud's model of the mind is built on the dynamic relations among the id, the ego, and the superego. The id includes internal drives and the need to satisfy them. Among the most general drives are the pleasure principle and aggressive instincts. The ego includes learned instrumentalities for satisfying these urges: it furthers the aims of the id.[14] The superego refers to socially derived norms or the social facets of our personality: our conscience. The superego is instilled by a social organization, such as the family and school; it represses, displaces—or postpones the gratification of—many of our drives, which causes frustration. But these social institutions are based on displaced and sublimated psychic energy, particularly the energy from guilt over conflicts between fathers and sons. In other words, civilization is a result of psychic forces and may itself exhibit its own psyche and superego. Civilization's institutions are essential for the extraction of natural resources and the control of nature, but these institutions gain their powers by inhibiting our drives. Hence our ambivalent attitudes and even hostility toward both nature and civilization.

Biological nature ultimately is at the base of our psychological drives, but Freud's theory emphasizes the psychological forces that transform biological nature and lead to the development of civilization. Civilization, moreover, provides us with the technology and organization to dominate the rest of nature, so that nature can be made to nurture us. Indeed, the highest forms of civilization are measured by their power to control nature in order to make it more productive for humankind.[15]

The Realm of Nature

Claims about the power of the natural world to determine human behavior include theories that reduce (in the technical sense) meaning and social relations to the forces recognized by the natural sciences. This reduction places natural forces in a position to determine the conditions of the other realms; it also presents a view of the influence of humans on the nonhuman realm as simply a subsystem affecting a larger system. I first consider theories that see nature as driving human action; I then turn to those dealing with the effect of human activities, conceived in natural terms, on nature; finally, I examine human ecology as a bridge between natural and social science theories.

The entire fields of neurophysiology and sociobiology, and large portions of genetics, are dedicated to the reduction of portions of human behavior to biological, chemical, and physical processes. Links have been established between specific genes and certain types of behavior; between certain chemical and electrical states of the brain and certain mental dispositions and activities. Assertions have been made that attitudes and values, social organizations and social hierarchies, territorial behavior, and the like are structured by our biological instincts and drives. Some researchers even claim they could predict which relative you would most likely sacrifice yourself for by considering which relative would perpetuate the most of your genetic pool.[16]

Other natural sciences focus on climate and the biosphere as the mechanisms that drive human behavior. Perhaps the most comprehensive environmental theory, which still has residual effects, is the ancient, and now discredited, doctrine of elements and humors. Through the correspondences of the elements of air, water, earth, and fire with the humors of phlegm, bile, black bile, and blood, the theory links natural forces that originate in the stars and planets with mental and social behavior.[17] The positions of the planets and stars were thought to affect the distribution of elements, which in turn affects the internal balance of the body through the distribution of humors, and this determines mental and social states and physical well-being. More recent theories of environmental determinism are far less sweeping. Elseworth Huntington views the local environment as a determining factor and supports his claims with specific relations, such as associations between temperature and barometric pressure and the expansion of the Mongols, and between temperature extremes and higher forms of civilization.[18] Such sweeping correlations have been subsequently tempered by environmental probabilism and possibilism, but these, as Roy Ellen points out, are still deterministic at heart.[19]

These theories emphasize nature's constraints on human behavior. They can make room for randomness and incompleteness, but there is no room for the individual as a free agent. Thus the natural realm views our transformations as inevitable, because we act according to our nature. If this means fouling our nests or depleting our resources, then that is the way it is. And even then, we may survive. After all, we have so far.

We are increasing in number; we continue to settle in practically every nook and cranny of the globe; and we are even thinking of colonizing outer space. Such implications of inevitability make many social scientists recoil from social hypotheses drawn from the realm of nature.

The discussion has so far focused on theories in which the forces of nature, through some kind of causal chain, affect the realms of meaning and social relations. But the same chain can be used in the opposite direction, by analyzing the human impact on nature by considering humans as subsystems within the larger natural realm. This means examining the physical, chemical, and biological outputs of human behavior without looking to the social or mental realms for the sources of such output, much as one would examine the outputs of volcanos without knowing why they erupted. Such an approach is essential in order to find out what humans in fact are doing to the natural world. But without explaining this output by means of theories from the realms of meaning or social relations, our understanding is restricted to what we are doing to nature and not why we are doing it.

One variant of seeing human actions as part of nature is phrased in terms of function and purpose: it argues that it is natural that humans, like any other species, try to survive. If in so doing, humans have altered and even simplified the natural environment, that is just the way it is. We may even so continue to succeed. This success might be assured because we do what is in our nature to do, or because the rest of nature is designed to accommodate us. The idea of a forgiving natural realm does not necessarily lead to a belief in biblical design.[20] Rather, it can lead to the byways of functional and teleological systems. Jim Lovelock, for one, in his gaia hypothesis, has argued that much of ecology fits into place scientifically if the earth is thought of as a living organism.[21] An extension of this argument could lead to the belief that the biosphere is designed to support human life.

A more balanced description of the complex links between people and nature is provided by human ecology. Ecology and ecosystems stress connectivity and mutual causality among the natural and human components. "In the ecosystem view," according to Ellen, "all social activities impinge directly or indirectly on ecological processes and are themselves affected by these same processes. The approach thus emphasizes the two-way character of causality . . . although the relative influence in reciprocally causal relationships is never equal and may be very unequal."[22]

I return to the issue of relative influence shortly; I want to first note that the primary, though by no means only, device ecologists employ to connect human and natural activities is, to put it positively, to focus on characteristics that both systems share, or to put it slightly negatively, to reduce human actions to physical ones. One of the most important and elemental connections is the flow of energy. Positing a web of energy and material relations allows the ecosystems approach to draw together natural and human processes. This focus specifies how nature and society are related: for example, how changes in irrigation practices (along with the human energy required to initiate and sustain them) change the caloric yield of crops, and how changes in precipitation also affect these yields. The analysis can be extended to the inorganic. It can help us trace the effect of agricultural practices on erosion and the effect of this erosion on stream morphology and flooding. It can be used to trace the effects of effluents throughout the material and biotic systems, if energy and material flows are known.

But what causes these flows? It is at this point that the single web of people and nature comes unraveled by the theoretical pulls from the forces of the three realms. An ecological concept that provides one of these pulls is adaptation.[23] Whether defined in a strict biological sense or in a more flexible sense of general strategies for adaptation, the concept cannot escape its biological meaning. If strategic adaptation is defined as actions that help attain goals, the emphasis might shift from adaptation to strategy, but this only forestalls the problem because any and every action would become strategic and, thus, adaptive.[24]

We have to turn to the other realms to seek causes. (Ecologists are involved in this process when they argue over whether they should call themselves political, economic, or cultural ecologists.) Causes for human actions that are not from the natural realm are found in the realms of both meaning and social relations. I now turn to theories from the latter.

The Realm of Social Relations

A host of social science theories offer explanations of why humans act the way they do. Only a few focus directly on the transformation of nature by humans, although some at least imply a relation between social relations and nature and also between social relations and meaning. Some social science theories explore such quantitative aspects of human actions as energy expended and movement in space and time. These theories can be linked directly to the human ecological models and even drive them—but not necessarily in the direction of equilibrium or adaptation. Roy Ellen, for example, considers human society, along with other biological systems, in terms of energy production, utilization, and exchange, all of which provide "the material basis of human existence." Energy becomes the starting point for "a materialist explanation of human social relations and the history of these relations."[25] Where a materialist link leads to is discussed later. I first consider another chain, which can also create and direct flows of energy. This one, however, is forged by a more conventional social scientific perspective on human behavior—of humans as seekers of pleasure—and forms the basis of conventional social science theory.[26]

Conventional social science theory is built upon a political economic tradition that emphasizes the importance of individuals, from the exaltation of the individual in Renaissance thought, to the development of individual liberties in Enlightenment political theories and constitutions, to the development of neoclassical economics. Contemporary social science has crystallized these forces into propositions based on the assumption that people are motivated by self-interest to seek pleasure and avoid pain and that society is, or ought to be, structured to facilitate this pursuit.

The pursuit of pleasure can be quantified in terms of a scale like money, which allows social science models to specify how society is driven to multiply its surpluses, consume more energy, and transform nature.[27] From this perspective, the object of human behavior is to maximize income or wealth or to minimize costs. The pursuit of pleasure does not necessarily lead to a need for more pleasure. Some people certainly are satisfied with what they have. But certain social relations tend (some interpretations say "ought") to push us in the direction of needing more because of the way its parts are connected.[28] At this level of design, there are numerous subtheories within the major thesis of pleasure, each using different means to show how self-interest operates within a set of constraints.

Neoclassical economic theory, for example, presents the agent as free to pursue self-interest but also as propelled to do so by structures that make "more" more compelling than "less." For example, economies of scale, which in neoclassical economics result from specialization and division of labor, mean that more can be produced more cheaply. Firms take advantage of economies of scale in order to survive in the marketplace. This means that consumers are presented with even more to consume. The economy encourages an ideology of "more is better": more means success, more means a higher standard of living, more is progress. This ideology is, of course, also encouraged by advertising, which promotes the idea that more is better as much as it promotes the sale of a particular item. Of course, people may simply want more, even without these extra pushes.[29] Once set in motion, these factors reinforce one another and become embedded within institutions, each trying to maximize its own sphere of interests. Each of these institutions and their social relations affect the natural world in terms of consumption of raw materials and expenditures of energy. Which factors are significant—and how and why—are the issues that distinguish particular theories in the classical social science paradigm.

Conventional social science theories can also be linked to nature and meaning by considering ideas, values, and beliefs to be the products of social relations. This link is made by the sociology of knowledge and the social construction of reality, but it is common practice in everyday social science—when, for example, a connection is postulated between social and economic status and religious and political values.[30] These links assume that meaning, including attitudes and beliefs about nature and even the practice of science, are molded by social relations: the mind constructs nature and reality, but the mind itself is molded by society. Marxism argues that meaning is driven by social relations, although different ones from those discussed by conventional social science. Marxism also contends that social relations are linked to material processes that directly affect nature.

Although Marxism draws on the concept of self-interest and the pursuit of pleasure, it does so with a different twist. Marx tended not to isolate and abstract individuals conceptually and theorize about their interests apart from social contexts and constraints. Perhaps the most basic belief, and the one that makes Marxism a materialistic philosophy (and thus a potential bridge between nature and society), is that humans and nature are dialectically connected through labor. As all other living organisms, humans consume—and thus transform—nature. This is accomplished through labor, which is natural and yet which has particularly social qualities and which allows us to eventually dominate nature. To paraphrase Alfred Schmidt: at bottom there exists only people and their labor on the one side, nature and its material on the other. People construct the world on the model of their contemporary struggle with nature. Historically, the struggle favors people.[31]

Labor, then, is both a link to nature and a means of transforming it. Labor is elaborated through production, which includes both forces and relations. The forces of production are resources and technology; the relations of production are the social organization of work and the ownership of the means of production and of surplus. Historically, there have been different modes of production: primitive, slave, feudal, and capitalist. The number and characteristics of these modes of production are unsettled issues in Marxist theory.[32] These modes affect material production, the use of nature, and the distribution of wealth. They also provide the basic organizing principle for other social organizations (such as education, child rearing, and leisure) and the base (in vulgar materialistic terms) for the realm of meaning (or superstructure). Vulgar materialists would say that the realm of meaning, or the superstructure, is an epiphenomenon reducible to the mode of production.

Because Marxism is a form of materialism, it can be argued that its proper location is within the realm of nature: materialism can be interpreted in a way that is close to environmental determinism.[33] However, this interpretation of materialism is not a primary characteristic of Marxism, because most Marxists see human behavior as a struggle to overcome and transform nature (we can quibble about whether this is itself natural) and as driven by forces in the realm of social relations. (In capitalism, these forces are class relations. Capitalism is unique in the scope and scale in which it has transformed nature.) Within the biosphere, there is virtually no nature that is not in some way affected by labor.[34] According to Marx, we never escape from the necessity of labor and thus the transformation of nature.[35]

Just as some Marxists interpret materialism so literally that nature appears to be the determining force, other Marxists emphasize the relative autonomy of meaning and see that meaning affecting social relations. Meaning is then not a superstructure but an integral component in the production of society. Yet even these Marxists agree that, in the last instance, the realm of meaning is driven by social relations.

Clearly, Marxism, like classical social science, is not monolithic. A narrow Marxism based on mode of production and class conflict as the necessary and sufficient social forces may appear logically coherent but is unrealistic. A broader interpretation that attributes power to other forms of social relations and to other realms faces the problem of logical incoherence. In addition, Marxist theory, like all other theories of human behavior, has difficulty finding room for free agency. Marx wrote that "men make their own history."[36] Yet an analytical perspective combined with the power of specific structures, especially class relations, are emphasized to such a degree that they cast the freedom of the agent into doubt.

It is important to note that technology, bureaucracy, and population pressure are parts of social relations and can be thought of as causing environmental transformation. Ever since Malthus, important social theories have viewed population and technology as major transforming forces. Both classical social science and Marxism recognize that demography is important, that institutions tend to have a life of their own, and that technology can lead to alienating consequences. But most social theories do not see technology, bureaucracy, and population pressure as root causes of our transformation of nature but as particulars of social relations that are driven by deeper social structures.

•

At this point, we have come full circle. All of the theorists discussed would probably agree that people and nature are related through flows of mass and energy, and they might also agree that space is an essential element in this connection. But beyond that, they seem to be deadlocked as to the relative importance of the three realms, overall, and in particular contexts. Some theories claim that meaning shapes nature and social relations; others claim that nature shapes meaning and social relations; still others that social relations shape meaning and nature. This circularity suggests that none of the theories provides a balanced picture of the center.

The problem runs even deeper, for it raises once again the relation between forces and perspectives—and this time in at least two senses. First, the relation surfaces because what we know about forces comes from theories that contain perspectives. This particular issue leads directly to the problems of realism, idealism, and skepticism. Second, it comes to the fore in the issue of agency—but this time in the sense of reflexive people who can themselves theorize about their own actions.

Forces and Perspectives

The foregoing partial and competing theories raise fundamental doubts about the possibility of objective knowledge. A profound relativity pervades the realms of meaning and social relations and even extends to the natural sciences through the thesis that reality is mentally and socially constructed. Contemporary philosophy and the history of science have been unable to demonstrate how it is possible for science to be objective, given the fact that models and theories are human creations; but these problems (perhaps fortunately) have not affected the everyday conduct of the natural sciences. Even though philosophers cannot show us how particular theories are more valid than others, science still seems to work. Its structure confidently supports or rejects statements about reality that are far removed from everyday experience. But the fact that we do not have a clear understanding of how our thoughts and perspectives about the world mesh with the "true" character of the world leads us to the issues of realism, idealism, and skepticism.

Realism, Idealism, and Skepticism

A very general form of realism is the belief that "the world is independent of our minds."[37] An equally general form of idealism is the claim that whatever exists must coincide with what we can sense and conceive of and thus must be something for which we can have evidence.[38] Neither position can ultimately be proved as true, and we must, and in fact do, draw on both. Idealism's strong point is its claims that we know the world only through our minds and that if there is a world out there that cannot be sensed or conceived of we will never be able to know it in any form. But doubting the existence of this unsensed world leaves the door open for us to doubt the existence of things that are as yet unsensed, and this could slip into a doubt about the existence of anything beyond our own personal and direct sensations, which would lead to the radical idealist or solipsist position of being locked within a world that extends no farther than our own minds. Certainly no sane person can adhere to this. Sanity requires the realist belief that there is a world out there that exists independently of us and even one that is (possibly forever) beyond our experiences and conceptions. Such a realist position cannot be supported by logic alone—it must simply be assumed.[39] The belief in the existence of a world independent of the mind, though we may never know that world as it really is, is one form of realism. Doubting its existence because we can never know that world as it really is, but only as we construe it, is a form of idealism. And the spark set off from the irreconcilable tensions between the two is skepticism.

The deadlock among the theories is due partially to these irreconcilable positions. But does that mean that particular realms imply one or another of these philosophical positions? At first glance, it might appear so. Those who attempt to understand the power of the mind might be thought of as advocates of idealism. For example, intellectual historians and philosophers of science remind us that what we know of nature comes from natural science models and hypotheses. These are languages that talk about and represent nature. We cannot know nature or anything else without representing it in thought, and to this extent the mind can be said to fashion the natural world—the natural world is in some sense a mental construct. This, however, does not necessarily support the claim that there is no world out there independent of the mind. It simply means what it says: that we know the world only through our attitudes and beliefs. The realm of meaning thus does not in principle commit us to idealism. In fact, all of the theories about meaning examined above are realist theories in that they claim to disclose forces or properties of thought that exist even if we are unaware of them (or that, if sensed at all, would be sensed as autogenic action). Furthermore, realist theories also maintain that there could well be other, as yet undisclosed, forces at work.

By the same token, natural scientists who explore the world beyond us—the world experienced as impact—might appear to be committed to realism, but it is possible for them to hold an idealist position and claim that there is no external reality. Instead, all that exist are our models and theories. The fact that one or another of these realms in themselves does not imply an idealist or a realist position means that we must look elsewhere to understand how these issues affect the deadlock. This means that we must turn to two related problems. One is the fact that our knowledge of the forces is based on theories that themselves contain perspectives; and the other is that, in the social sciences, we are both observers of the world and a part of that world: we are both subject and object.

Reflexivity

I noted that perspectives, such as science and religion, can also be forces, since they can have powerful effects on our behavior. Indeed, any system of meanings can be considered a perspective if it is comprehensive and self-critical enough. All of our knowledge about forces is just that—knowledge. That means that, even though we may believe that this or that force really exists, we cannot escape the fact that we know of it through our theories and models. If we dismiss these theories because they are simply perspectives, we will implicitly embrace idealism. We must make the realist leap of faith and assume that the theories and models, while not necessarily identifying things as they really are, nevertheless provide the basis of what is known about the world. Our acceptance of both realism and idealism does not, however, help us decide what is a true or false impression of the world—but neither does our acceptance of idealism or realism separately. Each on its own leads to equally inconclusive positions about verification; neither can resolve the problem of how our ideas correspond to reality.

That problem may never be solvable philosophically, yet it is constantly overcome in our everyday lives when we operate as our own theoreticians, explaining the world we inhabit and our own behavior in that world. Acting reflexively is related to our freedom as agents. Free agency becomes ever more difficult to discern as we move toward a view from nowhere. Since most of our knowledge about forces comes from theories that assume an objective position and possess an obvious bias toward causal relations, these theories have difficulty addressing the realm of free agency. Social science need not express these forces as causal laws. In fact, in the social realm, forces appear most often in the form of rules and regulations over which we seem to have little or no control.

Still, most of us believe that our sense of free agency is not an illusion. Even though our actions are constrained by rules or laws, we are still able to create projects. We set our own goals and attempt to attain them.

We provide our own explanations for human actions, and we learn, evaluate, and react to the theories or explanations of others. In all of these respects, we are free and also reflexive. Thus, unlike natural sciences, where subjects study objects, the social sciences have the problem of subjects studying subjects. This reflexivity means that knowledge of theory can alter our own behavior: that theory transforms its own object.[40] From its very beginnings, social science has held this view as well as its more conventional and less reflexive views that are closer to nowhere, and it has developed attendant methodologies, including *verstehen*, empathetic understanding, and other humanistic approaches, which are quite different from those of the natural sciences.[41] These methods attempt to generate theory by examining behavior from a position closer to somewhere.

Reflexivity complicates the already thorny issue of verification. Evidence is extremely important in a society committed to science, and people who, in their everyday lives, theorize about their own behavior and the behavior of others must have their theories conform to their experiences. But many of the concepts and posited structures of social theory seem remote from our day-to-day experiences. When people can provide their own reasons for their own behavior, they may be skeptical of social theories that postulate hidden forces and undisclosed meanings. Because of social science's subject/subject relationship, it matters that social theories are remote from everyday experiences, and this must be taken into account when evaluating social theory. Reflexivity does not make objectivity evaporate altogether, but it does make it more difficult to grasp.[42]

If we de-emphasize reflexivity and free agency, our behavior will be both *predictable* and beyond our control. If we emphasize reflexivity and free agency our behavior will be *unpredictable* but within our control. If these oppositions are not bridgeable in theory, they are in everyday life—or at least our day-to-day actions provide the illusion of such a bridge. This, in no small measure, is due to the integration of forces and perspectives within space and place, which leads us back to the geographical premise.

Space and Place as Forces

The geographical premise is that space and place are fundamental means by which we make sense of the world and through which we act. Does this mean that space and place are forces? This question is basic to geographical theory. At the level of everyday life, the answer is self-evidently yes, which is why geographical inquiry is vital to understanding our place in the world. (This understanding would develop even without the assistance of professional geography, although slowly.) Space and place are powerful forces in human experience and they are essential aspects of being in the world. From this perspective, the sense of personal place that is rich and complex and possesses something of a holistic quality seems to take primacy over the sense of space: from the personal perspective, space is little more than a sequence of places.

But the question is not often formulated from the personal perspective of daily life. Rather the question about space and place as forces has usually been asked from more distant perspectives, which then view place as simply the location of things in space. From these perspectives, personal place becomes vague and all but disappears. This means that, in the following discussion, space supersedes place in importance (except in the realm of social relations, where place reasserts itself through territoriality), so that the basic question becomes how space is a force within each realm. The role of space may not be clear to those who practice in only the realms of social relations or meaning, but it is clear to those in the realm of nature, which has always seen space as a fundamental property of the world.

Theories from meaning and social relations often neglect space and place and have been impoverished by this neglect. Several alternatives are available for rectifying this problem. The most common one is to insert space or spatial relationships within these theories. However, many of these theories are probably not worth salvaging precisely because their neglect of spatial relations has made them unrealistic. The second alternative is to put aside the neglect of spatial relations by particular theories and demonstrate instead how space is in fact an essential component of meaning and social relations and forms part of their ontologies. This is the approach sketched below. Although it can show how space is a basic component of all the realms and can redirect research within each to incorporate space, it suffers from the fact that these roles of space are not interrelated, because the realms are still fragmented. A third and most ambitious approach is to use space and place to reconceptualize the connections among the forces and perspectives. A full development of this alternative is far beyond the scope of this book, but portions of it are introduced in the discussion of morality. Consider, then, the second alternative: the role of space in the foundations of each realm.

The Realm of Nature

The natural sciences have long held space (and time) to be fundamental properties in the scheme of things. Spatial variables, such as distances and lengths, are prominent in all areas of science. Still, difficulties remain regarding the role of space as a causal agent; these difficulties center on the question of relative versus absolute space.[43] In physics, these terms have less to do with the perspective from which we view space than with whether space is a force that is immune to other forces. Absolute space means that the continuum called space is immune to influence, that its structure is rigid and cannot be changed by matter or energy, that its description in geometrical terms is independent of our viewpoint or frame of reference, and that these very features of space can exert physical effects.

Relative space, on the contrary, assumes that space can be acted upon, that its properties and descriptions are dependent on the distribution of mass and energy, and that, by itself, space therefore does not exert physical effects.[44] Newtonian physics tends to conceive of space in the absolute sense. With the advent of relativity, the relative view seemed to triumph over the absolute view, for the nature and form of space was to be everywhere dependent on the existence and distribution of matter. This conclusion seems to have been too hasty. As Einstein suggested, there has been no clear victory. The replacement of the absolute view with the relative view is still incomplete.[45] The dependence of space on matter and energy has found only limited expression in the general theory of relativity. While space is relative in the sense of its association with time in a space-time system, there seems to be as yet no way of completely eliminating the concept of absolute space from the equations of relativity. For instance, it is possible to speak of the shape of space when matter is absent, and it is necessary to assume boundary conditions about space to solve field equations.[46]

For earth processes, especially those of interest on the geographical scale, the nature of space should be described in a modified relativistic sense—or relationally.[47] *Relationally* means here that the properties of space are absolute and describable in Euclidean terms, because its structure is not influenced by the local distribution of mass; space exerts effect but not absolutely, because on earth space is not empty, and its effects are mediated by matter or substance. Thus the time it takes to travel depends not simply on distance but on distance modified by the mode or medium of transportation.

The relational concept shows how space is inextricably intertwined with substance at the most basic level as well as at the level of appearances, so that space enters into the equations of theoretical physics as well as those of climatology, geomorphology, and soil science. Such an embedding of space with substance does not make space an independent force but, rather, an equal partner in constituting force. It also makes clear that all interactions occur in and through this spatial manifold—

and thus are spatial—and yet the interactions must be constituted by chains of substances.

The Realm of Meaning

An extremely influential and comprehensive theory of the power of space in the realm of meaning is found in Kantianism.[48] This position argues that space and time, as a priori, are constitutive of thought and experience. According to Kant, space and time are conditions of awareness. As the only two pure forms of sensibility, they are necessary conditions for sense experience, and they order our senses in a space-time manifold.[49] Experience, and even awareness, to Kant are simultaneously and equally about ourselves and our world. "The consciousness of my own existence is at the same time an immediate consciousness of the existence of other things outside me."[50] A theory such as Piaget's, which draws on Kantianism, describes exactly how these stages of awareness develop and how, in the same instance, this means a development of space and time.[51]

If space and time are part of our intuition, does this mean that they do not exist in reality? In other words, is Kantianism necessarily an idealist philosophy? The answer is no. Kantianism does not deny the existence of a world as it really is. Moreover, Kantianism can be used to make a stronger claim about existence: space and time are real, and they can be used to define what is not real—as that which cannot appear within space and time.[52] Kantianism is a categorical statement about the basic constituents of the mind and has generated rich hypotheses concerning the ways in which space and time help constitute different types of meanings.[53] For example, different artistic forms or meanings result from different combinations of space and time. Science, ritual, myth and magic, and everyday life can be analyzed in terms of how they use spatial and temporal intuitions.[54]

The Realm of Social Relations

What of space in social relations? There are two related means by which space becomes an elemental part of this realm. The first is by adopting the relational concept of space. This enters the social realm simply because social relations are about individuals and their connections with one another through space and time.[55] Space provides the same function here as it does in the natural realm.

The second means of introducing space into social relations is through territoriality, the human strategy to affect, influence, and control things, social relations, and meanings by controlling area. We use territoriality to construct spaces and places as part of virtually any activity. The emphasis is on construct. Territoriality and the relational concept go hand in hand. Consider the possibility of an individual or group attempting to affect, influence or control the natural or the social environment. The relational concept argues that such attempts require the transmission through (nonempty) space of influence or power. Thus, if I wished to assure the safety of my young children in my own home by preventing them from touching this or that electrical appliance, I would need to be in contact with them to tell them or show them not to touch things or I would have to place these dangerous objects out of reach. These efforts must occur through space. But I can, and eventually do, supplement these spatial activities by the use of territoriality (another spatial activity) by declaring areas or rooms that contain these appliances to be off limits to them.

Territoriality is a spatial strategy to make places instruments of power. The relational concept and territoriality are about establishing accessibility, and each requires the other. Territoriality is an effective device, and often an essential one, because of its potential effects, which are outlined by the theory of territoriality.[56] In my example, setting the room off as a territory can work because of territoriality's ability to make it unnecessary to enumerate the things, social relations, or meanings we wish to control. Instead of stipulating that it is this or that phenomenon we do not wish to be accessible to others, control over the room settles the issue: the room, or territory, becomes the object of access. The room as a territory is embedded within a host of legal and customary territories that interact with one another, ranging from the use of the house as a private dwelling, to property rights, to state and national jurisdictions. These, in fact, may be so familiar that they come to be thought of as natural features of the landscape and not as instruments of power affecting accessibility to social relations and meaning and providing the geographical foundations for our day-to-day interactions.

The relational concept and territoriality provide our most basic understanding of how space helps constitute social relations. They support the general claims that place matters, that space is socially constructed, and that distance can have an effect on social interaction. They are crucial for understanding the particular geographical formulations of spatial analysis and the geographical attempts to make spatial relations necessary components of particular social theories.[57]

•

Space and place, then, are basic components of all three realms, exercising their powers differently in each, or sometimes jointly. It is extremely significant, though, that the penetration of space as a force is most clearly expressed in the realm of nature, where its role in theories is far more consistently maintained than in the realms of meaning and social relations. Geographers and others have attempted to correct this, with varying degrees of success, by reinserting space into particular social theories or by developing their own.[58] This neglect of space has contributed to both conceptual fragmentation and the impoverishment of theory. Such theories become too vague—not general or abstract, which are characteristics of the natural sciences, but vague because they cannot address causal chains through space and time. They are vague also because, by posing relations in such a way that they ignore the fundamental role of space and place in everyday experience, the statements these theories make have little correspondence to our own frames of reference—which, in the last analysis, are the principal means by which we judge the validity of social theories.

We are, after all, reflexive agents: we understand ourselves in terms of our own experiences. This leads to another complication stemming from the neglect of spatial relations. We cannot escape the sense of being in place, yet among the kinds of places we occupy most in the modern world are those of consumption; and these pose a fundamental geographical problem, because consumption creates places that have the appearance of being unconnected by anything else in space. This spatial dissociation or disorientation allows us to think we are both in place and in no place. In this way, our everyday experiences of place in the consumer's world do not provide a corrective to the contextless social theories. Indeed, they even encourage the illusion that things occur beyond the context of space and place.

· II ·

The Consumer's World

5
Place and Modern Culture

The intellectual world provides competing theories and perspectives but little common ground. This condition, not unexpectedly, is also found in modern culture, for culture can be thought of as the combination of perspectives and the forces of meaning and social relations in everyday life. How culture is created (or produced) and sustained is simply another way of asking how various aspects of meaning, social relations, and perspectives are connected, and this question leads to the same theoretical deadlock we have already encountered, a deadlock manifested in the confusion about and fragmentation of modern culture. For example, there is high culture, which emphasizes the power of meaning and its ability to drive social relations. There are popular culture and the culture of resistance, which emphasize the role of social relations in the production and circulation of culture, and which employ the concept of hegemony to characterize the subtle struggles among forces.[1] There is also the observation that, whatever the specific features of modern culture, culture seems fragmentary and transitory—a collage, a montage, a pastiche. This is the refrain of postmodernity.

Despite these tensions, modern culture still provides enough common ground for us to function, though one can say that what we have in common is more a society than a culture in the sense of deeply shared meanings. Modern culture comprehends large numbers of people and produces enormous quantities of conjoint activity. It does so in part by producing meanings that are thin enough (to borrow from Clifford Geertz's distinction between thick and thin descriptions) to be shared quickly by numerous people with differing and changing backgrounds.[2] This does not mean that our culture does not create profoundly complex and beautiful things. Instead, the meanings we share most as a culture are not generated by these creations but rather involve thinner meanings, which can be grasped quickly. In geographical terms, this is manifested in a shift to a more abstract and public sense of space. Regardless of which strand of modernity we emphasize—the future oriented and global, the nostalgic and local (which hopes to reinstate a thicker description), or the postmodern—these strands assume that modernity involves this thinning out of meanings, which helps distinguish our age from the premodern, a time when forces and perspectives were more tightly woven and when shared meanings resonated throughout the culture.

This comparison is at the basis of Émile Durkheim's characterization of the premodern as a society of "mechanical solidarity"—as opposed to modern society's "organic solidarity"; it is also found in Ferdinand Tonnies's distinction between *gemeinschaft* and *gesellschaft*.[3] And it pervades contemporary anthropology's assumption that in many premodern societies there was a profound and coherent structure that tied together the culture's systems of beliefs and social relations and that made it more static, or "cold," as opposed to the more dynamic, or "hot," modern society.[4] Certainly the past can be romanticized and the differences between then and now exaggerated, but the very condition of modernity assumes that such differences exist.

How does modern culture lead to a thinning out of meaning to provide a common ground? Places of consumption, of course, are the principle and most effective devices, but by no means the only ones, and to see how they work we must be able to compare and contrast their characteristics with, and their connections to, other examples in the modern landscape. A review of how modern culture creates shared meanings is beyond the scope of this discussion, as is even a comprehensive picture of the shared geography in the modern world. Rather, to provide a background for subsequent discussion of how places of consumption create contexts, I focus on three other means for attaining shared experiences in place: utopias and planned environments, nationalism and state territories, and mass communication and television.

Utopias attempt a comprehensive (or thick) integration of meaning, nature, and social relations, but they are shared by only a very small fraction of the society. By providing to these few a shared experience that is different from that of the culture at large, utopian communities even increase cultural fragmentation. Nationalism and state territories provide public sentiment about place, but because of modern skepticism and secularism, this feeling does not penetrate deeply into everyday life. And modern forms of mass communication, especially television, provide instant global access to place, but in so doing present them with little context and with thinned out meanings. Each of these is of course but a component of modern culture—with the most pivotal geographical element being places of consumption—and each influences the other. Before we examine them, we need again to set the problem by reentering the personal sense of place.

Personal Place

Although we have lost our innocence by tasting of the fruit of forces and perspectives, this awareness can be stayed for a time when we experience again the holistic sense of personal place. This is the preconscious sense of place that simply comes from being in the world. It does not separate the subjective from the objective, and it draws together forces from each of the realms. We could return to this experience by thinking of the restaurant, where we are relaxing in our chairs and letting our minds wander. At the same time, we are also the center of an environment that contains elements of nature in the temperature, pressure, humidity, and brightness of the room; elements of social relations, in that it is a workplace for the restaurant workers, a social setting for the customers, and a source of revenue for the local tax board; and elements of meaning, in that the restaurant symbolizes an emancipation from kitchen work and a place for some self-indulgence. Simply being and acting in this place ties these elements together.

Yet we live in a world that has prepared us to think along lines that unravel these connections and that lead to views that are somewhere or nowhere or lead to forces that spring from nature, meaning, or social relations. But we, ourselves, on the personal level, can still find our way back—literally or through imagination—to the integrative experience of being there. The return for each of us can partially stay this unraveling, but the very conditions of being in a dynamic society with an elaborate division of labor makes the contents of this sense of being in place difficult to share or even to communicate. Our feelings about this restaurant may not be the same as those of a more casual visitor and may be difficult to convey to people who have never even seen the place.

As we become conscious of the fact that most of us hail from different parts, do different things for a living, and adhere to different values and outlooks, we are less confident that the contents of the experiences of being in place can be communicated and shared. But the contents of our personal sense of place are not completely isolated from the rest of society. Place-creating or place-altering devices exist, such as utopias, nationalism, and the media. These devices attempt to bring aspects of the public and the private closer, to attain some degree of balance among nature, meaning, and social relations. These devices however, offer only provisional, temporary, and often local remedies and may have the long-run effect of exacerbating differences. Moreover, when these attempts are analyzed, we cannot help but see them from perspectives that our society has already developed—perspectives that emphasize once again the split between somewhere and nowhere and among meaning, nature, and social relations.

Utopias

Utopias are attempts to combine forces and perspectives in a shared place. They are extreme cases, because they advocate an abandonment of the larger society and the formation of new "cohesive wholes," in Yi-Fu Tuan's term.[5] Utopias offer totalizing solutions; they organize all of the significant facets of life, at least as utopian theorists define them. Among the earliest Western literary descriptions of utopias are the Eden of the Bible and the *Republic* of Plato. The tradition is continued in Thomas More's *Utopia,* Francis Bacon's *New Atlantis,* Campanella's *City of the Sun,* and in such contemporary writings as Wright's *Islandia.*[6] The term *utopia,* first used by More, means literally *no place,* but most of these idealized places are still imagined to be placeable somewhere on the earth—between Brazil and India, for More's Utopia; Sri Lanka, for Campanella's City of the Sun; the South Sea, for Bacon's Atlantis. Utopian literature provides blueprints for the harmonious integration of forces in place. These blueprints tend either to place enormous faith in the potentials of technology and complex social organizations to control and order nature or to distance themselves from technology and emphasize the past and the local in order to recapture the harmony and ease that might come from simplifying social relations and living in a nurturing environment.

Even though utopias are extreme and idealistic, they have inspired the formation of actual communities in real places, and herein lies the paradox. Utopias establish alternatives to the contemporary social order that are usually geographically far from the core of society. In the United States, for example, of the approximately 127 utopian communities established between 1663 and 1860, 11 were located in areas of fewer than six persons per square mile. The 116 communities in more densely settled areas still were, on average, only 190 miles from the frontier.[7] Although utopian communities are carefully planned, they never seem to sustain themselves over long periods or to be independent of the society they wish to reform. In this respect, they offer a shared and harmonious integration in place for only a few, who then are removed from the rest of society—which, by virtue of this utopia, may become even more fragmented.

Perhaps the most long-lived (but still dependent) attempts at creating "cohesive utopian wholes" are religious.[8] This is because those who join believe that the meaning of the community—its values and ideals—are not human or social creations but rather come from some transcendent power or god. These values are less likely to be subject to dispute or alteration. Secular utopian communities are less stable because they lack a transcendental authority: as human creations, they are as fallible as the humans that created them. According to Tuan, "their stability depend[s] much more on subjective good will and on shared ideals, the validity of which [does] not go beyond the dreams and desires of individuals."[9] Although the modern landscape is pockmarked by utopian attempts to mold harmonious connections among nature, meaning, and social relations, by their very nature, utopias cannot provide shared experiences of place for an entire society. Those who live in them might believe they work, but for the rest of society, which has no choice but to examine such places through its deadlocked theories and perspectives, utopias add yet another perspective and another experience that are difficult to integrate.

Less comprehensive planned environments exist, including new towns, new neighborhoods and work environments, and ecologically responsible communities. These, though, are not total environments, and they are certainly not based on ideas that have divine authority. As voluntary organizations or communities linked to society at large, they tend to emphasize one or another aspect of meaning, nature, and social relations and thus suffer from the tensions of the larger society and from the frictions and stresses of internal differences and dissent.

What can be said of the society at large? How does it attempt to share experiences of place? One of the most obvious ways is through nationalism.

Nationalism

Nationalism helps make place the focus of public sentiment. Political units, and especially modern nations, are fixed to geographical territories. Shared attachment and affection for the territory is encouraged and becomes inextricably mixed with loyalty to the state or the country, as the words *motherland* and *fatherland* suggest. In the United States, attachment to place has been encouraged through the use of ideas such as manifest destiny to describe the "natural and proper" extent of the nation; song lyrics, such as "O beautiful for spacious skies," to reinforce this sentiment; and national places and monuments to commemorate historic events of national import.

This connection of nation to place seems to include both generic and specific components. At the general level, it can be argued (from the theory of territoriality) that complex political organizations—confronted with often unpredictable tasks and requiring resources that cannot be enumerated—must use territoriality (not in the vague sense that they must occupy space or be in space, but in the precise sense of the definition "as the attempt to affect, influence and control people, phenomena, and relationships by delimiting and asserting control over a geographic area"), because it allows them to claim jurisdiction over anything and everything within their area. Territoriality thus provides the most efficient and—on a large scale—often the only means of controlling access to as yet unspecified things. This makes the relation between nation and territory essential. But it is a relation that does not single out any particular territory. In this sense, it is too general.

Real nations need a more specific connection to real places, because these help define the uniqueness of the nations and also help reify them. Plato recognized this problem (though in the context of the city-state) in his attempts to instill loyalty to his new and ideal state. He knew that abstract principles would not work by themselves. Rather, they would have to be abetted by that mode of thought that most effectively binds sentiment to a specific place—the mythical. The myth that Plato introduces is supposed to be communicated gradually,

> first to the rulers, then to the soldiers, and lastly to the people. They are to be told that their youth was a dream, and the education and training they received from us [the philosopher kings] an appearance only; in reality during all that time they were being formed and fed in the womb of the earth; where they themselves and their arms and appurtenances were manufactured; when they were completed, the earth, their mother, sent them up; and so, their country being their mother and also their nurse, they are bound to advise for her good, and to defend her against attacks, and her citizens they are to regard as children of the earth and their own brothers.[10]

A group's commitment to a particular place requires that it be infused with shared meanings, and the strongest means of doing this is through myth, ritual, or religion, because within this intense mode of experience, symbols become conflated with what they represent.[11] Mythical and religious experience are personal, but the issue here is whether the content of this experience can be shared and reinforced by the entire society, making it public as well as private. It is logical to assume that the best chances for this to occur are in societies that are static, closed, and do not contain conflicting views. This kind of a society reinforces a sense of shared experience and the notion that its symbols and forms of representation are not simply human constructions but are somehow natural and part of the larger system. Such societies are normally identified with premodern cultures, and the evidence shows that people in these cultures do indeed believe that specific geographical places, as well as forms and shapes in the landscape, contain enormous power, which resonates throughout the society.

The imperial Chinese city, for instance, was designed to reflect the form of the heavens, and each provincial city was to recreate this form.[12] These were not simply stylistic concerns. It mattered immensely, for the cities not only represented the heavens, they recreated and captured the power of the heavens and thus energized the lands. Such symbolic recreation of the sacred is common in premodern societies and penetrates the fabric of their cultures. It is understandable that places and geographical features possess these extraordinary properties, because premodern societies strongly believed they were under the control of natural forces, and their land alive with spirits.[13]

Modern society also attempts to infuse national place with something like a sacred meaning, as in our expressions of place as the fatherland or motherland. It does this on smaller scales through national parks, war memorials, and historical sites and museums that allow visitors to step back into the country's past and experience their roots.[14] I qualify the word *sacred* with "something like," because one of the characteristics of modernity is its combination of skepticism and secularism. This combination does not prohibit belief in a transcendent power; it does, however, make it less likely that there will be public consensus about the content and intensity of belief. Indeed, blind allegiance or faith in a doctrine, a country, a place, or anything, for that matter, would be regarded with considerable suspicion. Governments and society still attempt to impart this commitment, yet their efforts are diluted by secularism and skepticism.

This dilution is reflected in the contradictory terms, such as *secular ritual* and *civil religion,* that we use to describe shared beliefs.[15] These concepts draw attention to the important fact that civil and secular society does require commitment and belief in shared meanings and that perhaps the only means by which we feel confident that beliefs are important is to embed them in higher authority, or even in God. In the United States, civil religion, according to Robert Bellah, is not found in the "worship of the American nation but [rather in] an understanding of the American experience in the light of ultimate and universal reality."[16]

This ultimate reality implies a vague idea of a god who justifies the principles of American government. The Declaration of Independence, the Constitution, and other founding documents draw upon this transcendental source as their ultimate authority. The same appeal to a transcendent authority is found in the equation of important events in American history with events in the Bible.[17] The greatest American political rhetoric alludes to the War of Independence as a recreation of the Exodus of the Israelites, with Washington as Moses, and to the Civil War as a recreation of the death and resurrection of Jesus, with Abraham Lincoln as the savior. According to Ken Foote, secular ritual and civil religion mean that we transform historical sites and battlefields into national shrines to "celebrate and uphold the values and institutions of a secular cosmos" rather than sites that outline "a cosmo-magical myth of origins."[18]

But the transcendental background of civil religion and secular ritual is vague. It contains little more than the sense of something beyond ourselves that sanctions our views.[19] It is not specific enough to lead to particular claims on our behavior, and it certainly cannot lead to particular rituals. Moreover, even though some sense of a transcendental force may be shared by many, it faces an array of obstacles. In the case of the United States, state religion and blind allegiance, even to the government, are officially resisted by the same documents that make allusions to God. Our society is supposed to be secular, and the principles on which it stands were developed by particular individuals. These principles are social or human creations, and as such they encourage criticism and dissent. Calling the secular and the civil *sacred,* or joining them together in the awkward phrases of civil religion and secular ritual, obscures this important distinction.

A similar vagueness and ambivalence surrounds attempts to sanctify national places. In the United States, the obstacles to sanctification are many. One of them is the heterogeneity of its population, composed of immigrants and their descendants.[20] Another is that the Constitution does not conceive of the United States in any particular geographical form—its discussion of territory is abstract. The states that compose the nation are not named. The Constitution, in fact, contains explicit rules by which new states can be added and the boundaries of old ones changed.[21] The founding documents, in short, do not draw attention to the sacredness of place. The secular emphasis of modernity means that feelings about the sacred are difficult to officially sanction and reinforce. These difficulties create public skepticism about such symbols; this is especially clear geographically with war memorials and other national "shrines." Wars are especially revealing because of the principles they manifest. War memorials fix these principles on the landscape and demonstrate what parts of a country's past people want to remember and what parts they want to forget.[22] In the case of the United States, the striking aspect of its memorials is their incredible number and variety and the fact that there is no hierarchy or specified function associated with any of them. The public and the private, the sacred and the profane, are juxtaposed, so that almost anything counts as a commemoration of the past, and meaning becomes largely a personal matter.

Consider the public memorial commemorating Pearl Harbor. As with most events in history, the Japanese attack on Pearl Harbor is not easy to interpret. Its commemoration was delayed by the navy because it was seen as a symbol of defeat. For the navy, "honoring the dead [would be] the only acceptable reason for building a memorial."[23] But in addition to naval interests, there were the interests of "naval veterans, veterans of other armed services, the Territory (and later State) of Hawaii, the civilian survivors of those killed in the attack, the general public in Hawaii and on the mainland, and one or two commissions assigned the task of marking key wartime battles."[24] The result was a fairly antiseptic memorial to those who gave their lives during the attack. The monument was dedicated on Memorial Day rather than on the date of the attack.[25]

The events that modern national memorials commemorate and the decisions to commemorate them are created by people and are historically well documented and: they are not transcendent and immutable. The meaning of such places becomes even more mundane when we remember that most of the commemorations are not even national but are state, local, or even private. War memorials and commemorations can even be found in the home, as when space is set aside to commemorate the loss of a loved one or to display a collection of war memorabilia. Local private and public museums might commemorate wars by displaying war relics. Community buildings might commemorate wars by being named after battles, even though their functions may have nothing to do with the events. Even shopping malls and other commercial establishments might bear the name of such historical events.

The multifunctional character of some memorials makes them less than sacred. Even among nationally recognized memorials, there is no hierarchy of importance, as there would normally be among sacred places in an organized religion. And what power are these places supposed to possess? Certainly not the power of miracles or even of eliciting the truth (which was attributed to even minor Chinese city temples as late as the nineteenth century).[26] This does not mean that such memorials do not work. They do evince shared and often strong sentiments in the form of common memories. But mostly they work by thinning out the meaning of the events and the place so that they can be shared quickly by a modern, heterogenous society. People visit them not only to remember but to quickly and vicariously experience adventures of the past. In this sense, national memorials are more like generic tourist attractions and theme parks than they are like shrines and sacred places.

Eviscerating the power of the sacred is part of the general modern tendency to thin out culture and homogenize modern places. This tendency is supported by several modern conditions. One is that the use of the public, objective, geometrical meaning of space makes it difficult to convey the specific and emotional contents of place and thus tends naturally to emphasize their generic qualities. The same holds true of the scientific perspective.[27] Another modern condition is the trend toward a global economy and culture, which seems to require that places all over the world contain similar or functionally related activities and that geographical differences or variations that interfere with these interactions be reduced. After all, if we live in a global village, then we must feel at home anywhere, and the simplest way of making us comfortable is to remove the strange and the unexpected. The thinning out and homogenizing of culture can be expected as a consequence of yet another important condition—modern mass communication, especially television.

Place and Mass Media

Our experiences of the world are always mediated by culture. Even those sensations that seem a direct result of the proximate environment are selected and transformed by our feelings, tastes, and symbolic systems. They are also transformed by the communications media, which both add to and filter these experiences. The term *media* can be used to describe the means by which we gain information through other than direct contact. Information that comes through the written word, radio, television, and cinema comprise the *media,* or more precisely, the mass media (although it is possible to broaden the term *media* to include anything that transmits meaning over distance, and this can include virtually all symbolic systems). Although media transmit information, they do so selectively and in a particular form, so that the information becomes transformed. An event written about in the newspaper appears differently on television. This is due not only to the obvious fact that television emphasizes the visual but also to the way television uses words and images—which, in turn, makes it different from cinema. Of course, certain events can be staged solely for the media, which makes them media events.

The media transmit information, but a great deal of insight can be gained by concentrating less on this fact and more on the related role that

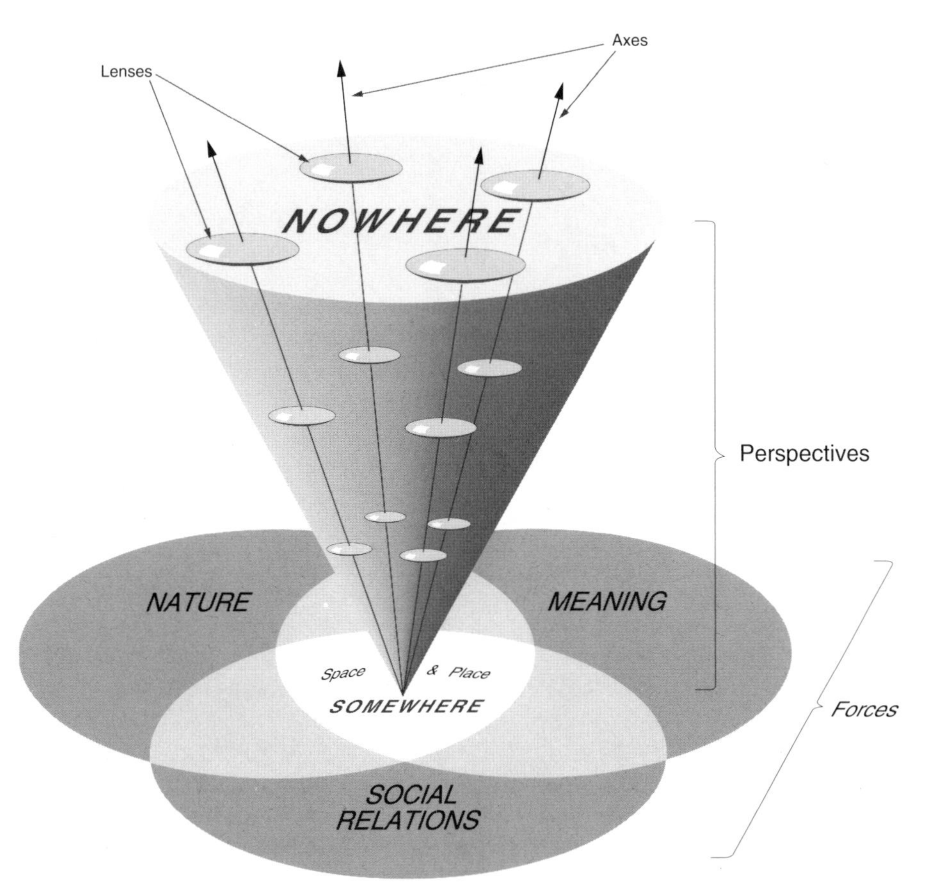

Figure 1. The Relational Geographical Framework. Space and place are central to modern concerns about the relations among forces (ontology) and perspectives (epistemology). Space and place help constitute and integrate forces from the realms of nature, meaning, and social relations. Place, having an inside and an outside, invokes perspectives ranging from somewhere to virtually nowhere. The axes between somewhere and nowhere represent particular forms of abstraction, and the lenses represent perspectives along these axes.

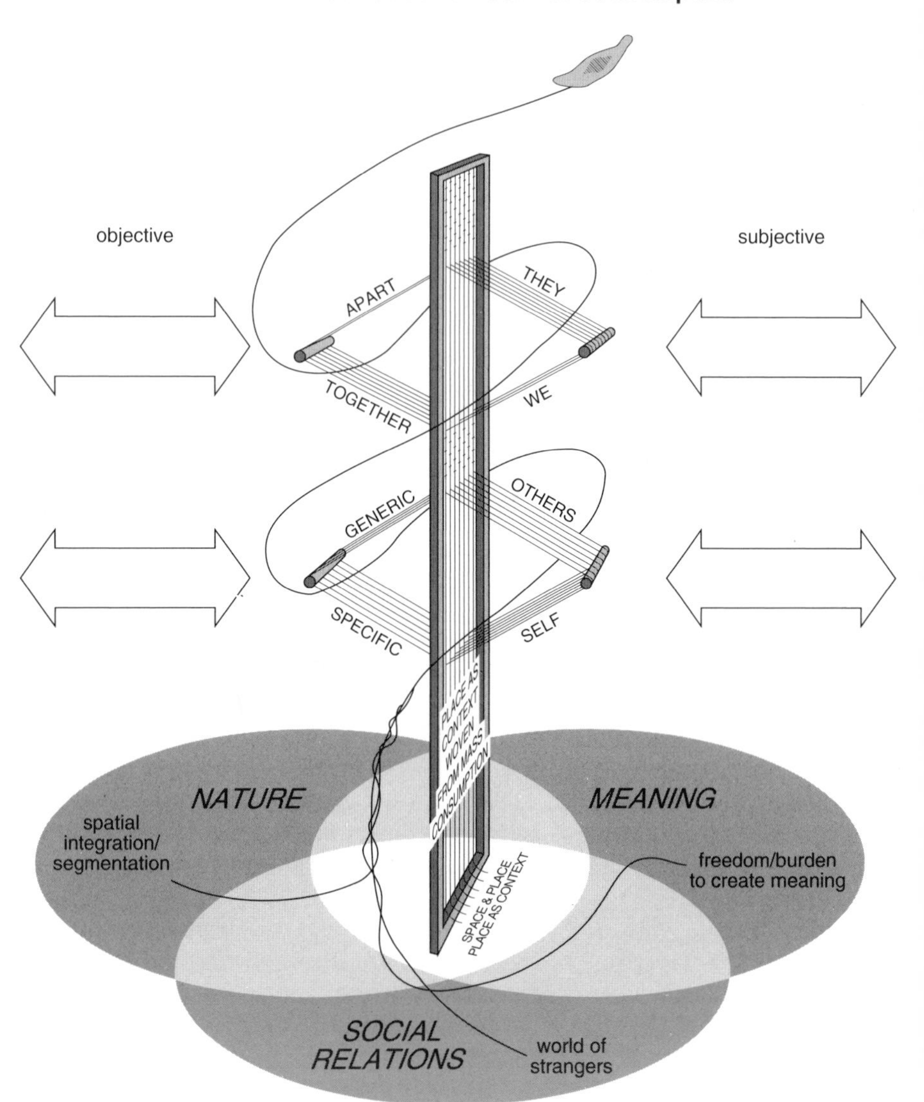

Figure 2. The Geographical Model of Consumption: The Loom. The relational geographical framework (figure 1) is reworked by the place-creating activities of everyday life. Mass consumption is the most powerful and accessible of these. It allows us to weave together elements from forces and perspectives to create places as contexts, but these places contain the irreconcilable tensions (of generic/specific, apart/together, we/they, self/others) embodied in the loom. These tensions emerged with the development of mass production and mass consumption.

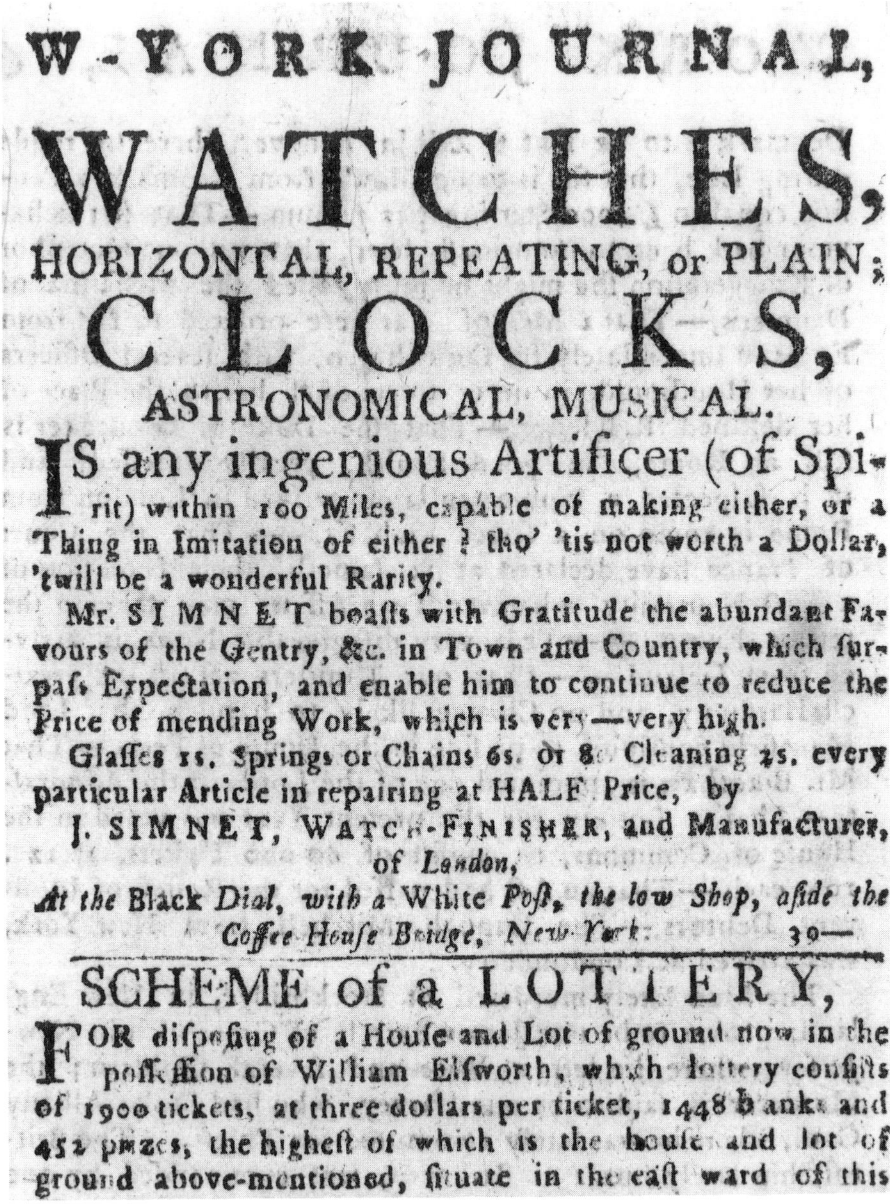

W-YORK JOURNAL,

WATCHES,

HORIZONTAL, REPEATING, or PLAIN;

CLOCKS,

ASTRONOMICAL, MUSICAL.

IS any ingenious Artificer (of Spirit) within 100 Miles, capable of making either, or a Thing in Imitation of either? tho' 'tis not worth a Dollar, 'twill be a wonderful Rarity.

Mr. SIMNET boaſts with Gratitude the abundant Favours of the Gentry, &c. in Town and Country, which ſurpaſs Expectation, and enable him to continue to reduce the Price of mending Work, which is very—very high.

Glaſſes 1s. Springs or Chains 6s. or 8s. Cleaning 2s. every particular Article in repairing at HALF Price, by

J. SIMNET, WATCH-FINISHER, and Manufacturer, of *London*,

At the Black *Dial, with a* White *Poſt, the low Shop, aſide the Coffee Houſe Bridge, New York.* 30—

SCHEME of a LOTTERY,

FOR diſpoſing of a Houſe and Lot of ground now in the poſſeſſion of William Elſworth, which lottery conſiſts of 1900 tickets, at three dollars per ticket, 1448 blanks and 452 prizes, the higheſt of which is the houſe and lot of ground above-mentioned, ſituate in the eaſt ward of this

Figure 3. A Craftsman's Advertisement. Before mass production and mass consumption, the tensions and contradictions (of figure 2) did not appear in ads. Rather, craftsmen and their shops were the objects of advertisements, as in this eighteenth-century ad.

56 HEAL AND SON, BED-ROOM FURNITURE AND BEDDING MANUFACTURERS,

OTTOMAN BEDSTEAD.

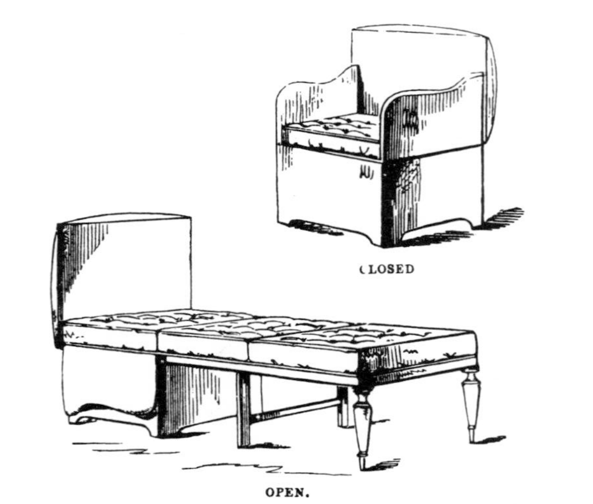

No. 123.

To be used as an Ottoman, Chair, or Bedstead, stuffed with woollen flock in printed canvas, sacking bottom, on castors.

Bedstead, 2 ft. by 5 ft. 8 in. £2 8

No. 124.

To be used as an Ottoman, Chair, or Bedstead, with lath bottom, stuffed with horse hair, covered in cotton damask.

Bedstead, 2 ft. by 5 ft. 8 in. £3 18

CHAIR BEDSTEAD.

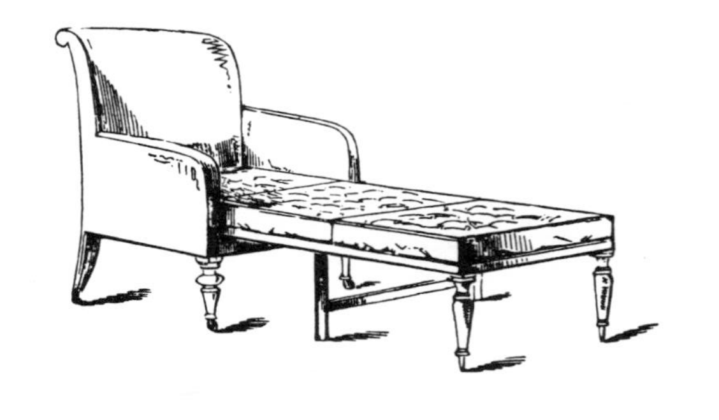

No. 125.

Chair Bedstead, stuffed with horse hair, covered in cotton damask, mahogany legs, French polished on brass socket castors.

Bedstead, 2 ft. by 5 ft. 8 in. £4 15 0

Figure 4. An Advertisement for Furniture. When mass-produced commodities entered the market, they needed no context. They were familiar products of obvious use. This ad is from the 1850s.

PAINTED WHITE, CREAM, OR ANY ART SHADE.

	£	s.	d.
3 ft. (**475**) Strong Iron French Bedstead, with brass top rods, and fitted with woven wire spring bottom	1	18	0
,, Good Brown Wool Mattress, extra thick	1	1	0
,, Second Grey Goose Bolster and Pillow	0	12	6

	£	s.	d.
Brought up	3	11	6
The Elton Suite, painted white, cream, or any art shade, consisting of 3 ft. Wardrobe with plate-glass door; 3 ft. Dressing Chest and Glass; 2 ft. 9 in. Washstand, marble top, tile back, and 2 brass towel rails; 2 Chairs	8	10	0

Figure 5. Furniture Advertisements that Create Context from Place. As more commodities swelled the marketplace and as social mobility increased, there was a need to make commodities more attractive and to specify what they could provide. Here, generic products create specific contexts by drawing things together. Although the tensions of apart/together, and generic/specific are present, the tensions of we/they and self/others are absent. In this period (the 1890s), it was clear which class we belonged to.

Figure 6. Jimmy Truck Advertisement. In the twentieth century, markets became flooded with products; social mobility increased. Products have to define themselves by creating not only a physical context, but a social one. Each ad becomes an idealized picture of how commodities empower us to create contexts, with ourselves at the center. This ad contains all four tensions equally.

igure 7. Royal Doulton China Advertisement. The juxtaposition of a dinner service with a landscape suggests that hina empowers us to create a romantic setting, one that differentiates us from other people. Other Royal Doulon ads show china in equally elegant settings, so that the entire ad campaign covers all of the tensions of the oom.

Figure 8. Howard Miller Clock Advertisement. A clock provides us with a world: if we purchase the clock, we "enter Howard Miller's world," which sets us apart from other people.

igure 9. American Airlines Advertisement. The generic becomes specific by drawing things apart and together in nusual contexts. Here, an airline becomes different by providing us and millions of other consumers with the ntire globe and, yet, in a way that sets us apart from others.

Figure 10. Canadian Club Whiskey Advertisement. An alcoholic beverage beckons us to enter its context and join the group. It separates one group from another.

Figure 11. Old Grand-Dad Whiskey Advertisement. An alcoholic beverage separates us from them.

Figure 12. Rodier Clothes Advertisement. Clothes help create a context, promising either to make us more like others or to set us apart. This ad claims that clothes, context, and consumer become one.

Figure 13. Claude Montana Clothes Advertisement. Fashion and place are both juxtaposed and interwoven.

Figure 14. Aramis Cologne Advertisement. Scent juxtaposes two worlds.

Figure 15. Chanel Perfume Advertisement, in Context. The ad admits that its context and claims are fantastic, which makes the product more alluring.

Figure 16. Chanel Perfume Advertisement, out of Context. We so widely accept that products empower us to create places or context that some products stand alone—no one context could do them justice.

*With permission from Ace Books. ©Infiniti Division of Nissan Motor Corporation in U.S.A.

CROSSING THE BOUNDARIES WHICH HAVE TRADITIONALLY SEPARATED THE DRIVER FROM HIS CAR.

The old experience is man-driving-car. The new experience is man and car driving.

There's a science fiction book called Hard-wired, where the author, Walter Jon Williams, talks about test pilots who plug into their aircraft through a super-advanced man-machine interface.

This is how he describes the sensation: *"(It was) a vision he could never share, never achieve anywhere else. A belonging, a completeness, that he could never talk about. Not even to those who flew with him. Just a shining in his eyes, a glow in his mind."**

We bring this up because it seems like a good way to help you understand "man and machine unity," an idea which pervades the Infiniti line of cars.

The Infiniti ideal is that the car should feel, in your hands, like a perfectly balanced tool. The power of the car shouldn't challenge you; it should enhance your ability by reacting predictably and easily to your natural movements.

To create this strong affinity (*affinity* not Infiniti) between the driver and his car, the unnecessary layers of high technology—so popular in this age of 'on-board' computers—were peeled away. Gauges are analog. Materials are traditional. Switches are designed with great regard for touch and feel. Technology is put to work where it works best. (In the suspension and in the transmission, for example: two areas where the application of technology can dramatically improve the driving experience.)

The attitude in the design of the driver's compartment is to make a place that feels comfortable and secure, but in touch with the car and with the exhilarating feel of the road. Behind the wheel, you should feel secure, relaxed and in control.

The technological underpinnings for the romantic notion described above are in rich array in the Infiniti line of cars. We suggest that you take a test drive.

For the name of the Infiniti dealer nearest you or for more information, call 1-800-826-6500.

Thank you.

created by Nissan

Figure 17. Infiniti Automobile Advertisement. Just as a commodity can appear without a context and still be understood to create a context, context can appear without the commodity.

I N F I N I T I

Figure 18. Advertisement for Bermuda. Tourist attractions exhibit the same tensions as other contexts created by mass consumption. Here, Bermuda sets us off from others (it is not meant for everyone). Place becomes us.

Figure 19. Advertisement for Michigan. Place and self merge.

Figure 20. Advertisement for Westin Resorts. Place not only defines self but separates us from them.

Figure 21. Advertisement for St. Croix, St. John, and St. Thomas Islands. Commoditized places become fantastic, just like mass-produced commodities.

Figure 22. Advertisement for Singapore. When places become tourist attractions and are mass consumed, they become indistinguishable from products. This ad makes visiting Singapore like shopping, allowing us to purchase the products of Singapore and, thus, the place itself.

Land . . . what else can give you so much pleasure now and for years to come?

The land we're offering is far from the crowds and rat race of cities, away from noise and pollution. It's nature in an unspoiled state.

Each ranch is 40 acres or more. A really big spread. With controlled access that assures exclusivity and privacy. We're up in the mountains of Colorado's glorious Sangre de Cristo range—the heart of the Rockies. Where deer, elk, eagles, wild turkey and other birds and animals still roam.

Very few owners will share this part of the American Alps, and our concept for buyers is simple: A large, desirable piece of property, offered with financing and full buyer protection. You can build here if and when you want. You may use it as a base for vacations, for cross-country skiing, hunting, fishing, hiking, camping and all kinds of outdoor sports and family fun.

It's the perfect place to acquire a substantial part of the American dream. Here you will taste life on the scale it was meant to be lived.

Forbes Magazine's division, Sangre de Cristo Ranches, put this project together based on the many requests received over the years for a really large tract of land. Through Forbes Wagon Creek Ranch, we're pleased to be able to share a part of it with you and your family. We've ranched this area for almost two decades and plan to be around for generations to come. Our neighboring Forbes Trinchera Ranch covers over 400 square miles, which is our firm commitment to the future of this unspoiled paradise in Colorado.

Ranches here start at $30,000. It's not a small sum. But unlike paintings and jewelry or new cars, this ownership extends past your lifetime and the lives of those you love to guarantee your own substantial heritage in America the beautiful.

For complete information, without obligation, call 719/379-3263 or write to: Errol Ryland, Manager, Forbes Wagon Creek Ranch, P.O. Box 303 C5 Ft. Garland, CO 81133.

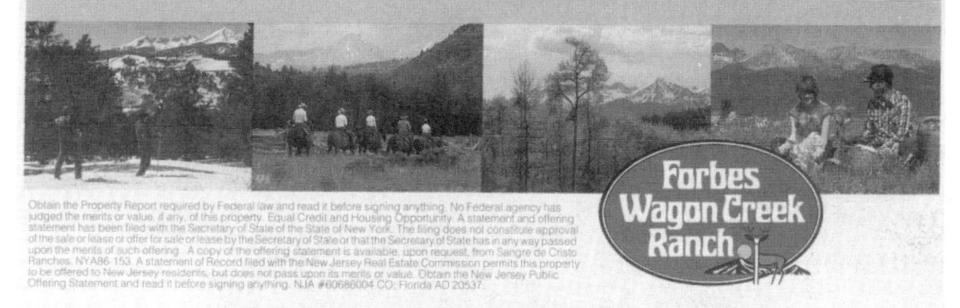

Obtain the Property Report required by Federal law and read it before signing anything. No Federal agency has judged the merits or value, if any, of this property. Equal Credit and Housing Opportunity. A statement and offering statement has been filed with the Secretary of State of the State of New York. The filing does not constitute approval of the sale or lease or offer for sale or lease by the Secretary of State or that the Secretary of State has in any way passed upon the merits of such offering. A copy of the offering statement is available, upon request, from Sangre de Cristo Ranches. NYA86-153. A statement of Record filed with the New Jersey Real Estate Commission permits this property to be offered to New Jersey residents, but does not pass upon its merits or value. Obtain the New Jersey Public Offering Statement and read it before signing anything. NJA #60686004 CO; Florida AD 20537.

Figure 23. Advertisement for Forbes Wagon Creek Ranch. A place and a view can be owned and framed.

Figure 24. SkyPager Advertisement. The juxtaposition of many contexts thins out meanings. Modern photographic and printing methods can make pastiches appear to be real—this realistic cityscape is nothing more than a composite of famous buildings. As a place, it is no place.

Figure 25. Ellis Island Foundation Fundraising Advertisement. The Statue of Liberty on a mountain range near a field of grain juxtaposes contexts, so that one place fits everywhere. Here, a society is advertised.

SHOW YOUR KIDS HOW A FOREST IS LOGGED.

Dress 'em up in old clothes and sturdy shoes, and we'll take you all "into the woods" on a comprehensive tour of one of our logging sites.

You'll discover how lumberjacks and foresters work with computer programmers and lab scientists to sustain our forests ***and*** their yields forever. Your kids will go "Wow!"

We offer you tours like this because we'd like you to understand that our mills and plants and logging sites serve your interests as well as ours.

And because we're proud of them.

For a list of tour locations, write Sharon Ramsey, Boise Cascade Corporation, One Jefferson Square, Boise, Idaho 83728.

Boise Cascade Corporation 

Paper/Office Products/Packaging/Building Products/Forests

Figure 26. Boise Cascade Advertisement. A commodity has replaced old-fashioned magic (sometimes magic is exactly what is being advertised).

Once you move in, it's not a house anymore...it's a home.

As a maker of home appliances, Whirlpool Corporation understands how important the home is to millions and millions of people. We know that it gives them a place of shelter, a source of pride and a sense of accomplishment. We also know that major home appliances go a long way toward making a house a home. This is why we promise to build and sell only good quality, honest appliances designed to give you your money's worth...and to stand behind them. We stand behind them by offering programs that include our Cool-Line® service, Tech-Care® service and helpful do-it-yourself repair manuals.

Standing behind our products is not just our way of doing business, it's our way of furthering the American dream.

Whirlpool
Home Appliances

Figure 27. Whirlpool Appliance Advertisement. An electric appliance transforms our houses into homes.

Figure 28. *Mother Jones* Advertisement. An ad advertises the problems of consumption.

ELEPHANTS STILL AREN'T OUT OF THE WOODS

We all know the African elephant is in trouble, but the trouble is under control, right? Wrong. The Humane Society of the United States has urged the government to declare the African elephant endangered; these magnificent beasts are still being slaughtered for their ivory tusks. Help keep the elephant on the plains of Africa—because they're still not out of the woods. Don't buy ivory.

REMEMBER THE ELEPHANTS . . . FORGET IVORY!

I won't forget the elephants. Please send me ______ "Remember the Elephants . . . Forget Ivory" bumper stickers at $ .50 each. I enclose an additional, tax-deductible contribution of $ ______. All items must be prepaid. Make all checks payable to **The Humane Society of the United States, 2100 L Street, NW, Washington, DC 20037.** Allow 4–6 weeks for delivery. We ship UPS; please include street address.

Name ______

Address ______

City ______ State ______

Zip ______ **M23**

Figure 29. Humane Society Advertisement. Advertisements of environmental concerns are subject to the same tensions and oppositions as ads that sell products.

Figure 30. PP Stocking Advertisement. An ad can be changed through graffiti and thus made into an instrument of resistance. Here, graffitilike words subvert the we/they tension in the ad by interposing its own definition of *we*.

igure 31. Bon Marché Exterior. From the beginning of mass consumption, places of consumption have themelves been advertisements. Bon Marché, the world's first major department store, was a vast display of goods nd contexts. It predates modern advertising.

igure 32. Bon Marché Interior.

Figure 33. Shopping Mall Interior. The modern shopping mall not only is a context for products, it simulates beaches and plazas, which are then sold as contexts.

Figure 34. Shopping Mall Interior.

INSIDE
WASHINGTON

Only those who know Washington inside out know about the Georgetown courtyard we've recreated under a glass dome. And about our exceptional service. All just 4 blocks from the White House and near the Capitol. Call your travel agent, Hilton Reservation Service or call us at (202) 429-1700.

VISTA
INTERNATIONAL HOTEL
WASHINGTON, D.C.

Figure 35. Vista International Hotel Advertisement.

media play in creating and transforming images of contexts, environments, and, quite literally, places.[28] Places, whether real or staged, are contexts for the information on television screens, and these contexts become part of the television viewer's personal environment: places on the screen become part of our living rooms and have an effect as real as the effect of their furnishings. Mass media researchers recognize this effect of television when they describe it as an information environment, symbolic environment, or an environment of symbols.[29] (Others make the even stronger claim that the "media are really environments" and that we live in our media.[30] This, I think, is going too far and is the same as saying that a book is a place when it becomes so absorbing that we lose ourselves in it; a place, after all, must not only have location (although this can move) in space, even if its extent can vary almost indefinitely, but also be literally possible to enter. This aside, it is clear that television has powerful effects on geography. Marshall McLuhan argues that

> electric circuitry has overthrown the regime of "time" and "space" and pours upon us instantly and continuously the concerns of all other men. It has reconstituted dialogue on a global scale. Its message is Total Change, ending psychic, social, economic, and political parochialism. The old civic, state, and national groupings have become unworkable. Nothing can be further from the spirit of the new technology than "a place for everything and everything in its place." You can't go home again.[31]

Gary Gumpert states that "when someone can be intimate with you [via electronic communication] and yet be physically so far away, the concept of the geographical universe is altered. . . . Auditory and visual images can transcend time and space."[32] The effect of mass communication on place must be analyzed through more than one medium; as Gumpert points out, "one medium does not constitute the 'media environment'; rather, that environment is composed of all media that are *potentially* accessible to the inhabitants of a geographical area, thereby providing a common pool of mediated experience."[33]

Still, it seems that television is the most powerful medium of all, and though its effects may not be shared by all the mass media, it has complex and important geographical consequences. One of these consequences is its tendency to thin out and homogenize the meanings of place. The view that television (and other electronic forms of communication) is a homogenizing medium (although it has many other effects, some of which counter homogenization) is implied by McLuhan and is summed up by Alan Rubin, Elizabeth Perse, and Donald Taylor: "The more time people spend living in the world of television, the more similar [become] their perceptions of social reality to television depictions."[34] These views are shared by Edward Relph, who writes that "mass media conveniently provide simplified . . . identities for places . . . and hence tend to fabricate a pseudo-world of pseudo-places," which helps create "a growing uniformity of landscape and a lessening diversity of places by encouraging and transmitting general standardised tastes and fashions."[35]

How does television tend to make places seem more alike? The process is complex but it can be simplified by considering it as involving three levels. First, the programs themselves, with their particular scenes, can be watched by almost anyone, anywhere in the world, at the same moment. This means that hundreds of millions of people in hundreds of millions of places experience the same things at the same time. It is as though we were there together, at the same place—but even more so, because we are all viewing things from the same camera angle and perspective, which would not be possible if we were literally there.

This may seem to argue for a thickening of our shared experiences, for it can be interpreted as melding a public sense of place with a personal one. But we must remember that we are not literally there, and thus we cannot affix other experiences to that place in order to deepen its meaning. The use of place as a scene or a setting in television plus television's reliance on the sense of sight tear place from its web of interconnections and insert it in transitory contexts. By the same token, if hundreds of millions of us in hundreds of millions of different locations spend hours a day sharing the same scenes and events, we are being virtually transported to the place portrayed, and the places portrayed are virtually entering and transforming the hundreds of millions of places we occupy. Fragments of Rome, Paris, Dallas, and the Sahara become parts of our living rooms. This makes our experiences and the contents of our homes more homogeneous. (We might feel that our environment is made more varied by television, but other people's environments would be changed in the same way.)

This fragmenting and thinning of the meaning of place is reinforced at a second level by the way television scenes and programs are arranged and ordered and by the relatively passive way we watch television. (I say "relatively" passive, because certainly we can be intellectually challenged by television or we can view it critically. McLuhan and John Fiske say we should not underestimate the depth of involvement it promotes.[36] But television seems to engage fewer critical faculties than do reading and conversation.) Watching television is certainly more passive than actually traveling to places, despite the efforts of tourism to make travel much like viewing television. Television uproots and juxtaposes one context after another far more than other media, and this obscures the real historical and geographical depth of place and weakens its relation in space and time to other places and events. Different contexts flash past by the mere changing of channels, each offering a string of programs in different settings.

Television's juxtaposition of places thins out their meanings and contexts and distorts them. Although it is often assumed that television is a "transparent" medium, providing visual fidelity of what is really out there, this is hardly the case. Television possesses its own filters, such as camera angle, film splicing, scene selection, dialogue, background music, to say nothing of the director's style and the program's intent, that distort even its attempts (through a *Nova* program or a television documentary by Granada) at rendering a real landscape.[37] Even television news has employed simulation.[38]

A third level at which television works to make places more alike has been explored by Joshua Meyrowitz. Meyrowitz recognizes that television thins out contexts, but he also draws attention to the fact that television alters social roles and makes them more generic. He claims that "by bringing many different types of people to the same 'place,' electronic media have fostered a blurring of many formerly distinct social roles. Electronic media affect us, then, not primarily through their content, but by changing the 'situational geography' of social life."[39] To understand how this blurring occurs, Meyrowitz joins the approaches of Erving Goffman and McLuhan to develop a framework for analyzing human behavior and, especially, television's contexts in terms of a "front," "middle," and "back" stage.

Any human activity, whether work or play, can be thought of as a production and, like theatrical production, involves support, training or rehearsal, and performance. All but the last—the visible performance—take place behind the scenes, or backstage. The actual performance occurs front stage. A university lecture is a front-stage activity, which requires backstage preparation of thought and research (and further backstages of education and training).

Front stage and backstage in this context are a continuum; the more (or less) formal an event, the more (or less) distance exists between front stage and backstage. At a formal dinner party, the guests usually do not see the kitchen, which is part of the backstage. At a more casual dinner gathering, the guests may gather in the kitchen and even help prepare the meal, which could even be consumed in the kitchen. We can say of the more casual example either that there is little distance between front stage and backstage (although one can stretch the continuum indefinitely by adding prior backstage activities, such as the host's storage of previously prepared food, and even the backstage support of grocery stores and food industries); or we can say that the event takes place in a middle region. This middle region is neither as formal as the dining room dinner party or as informal as the backstage might be if no guests were invited to dinner. This middle region blurs the distinction between public and private and also makes the activities blander and more acceptable to a more general audience.

Meyrowitz identifies this middle-region effect with television.[40] This region forms the third level of homogenization. Television is all-intrusive and ubiquitous: it penetrates everything and everywhere, so that "by revealing previously backstage areas to audiences," and yet staging this revelation so that it becomes a middle region, the context of television blurs distinctions between backstage and front stage and homogenizes social roles. Television's penetration into backstage and its ability to present backstage to practically anyone, anywhere "leads to the dilution of traditional group behaviors and the development of 'middle region' compromise behavior patterns." This means that "television content has evolved the same way that the conversations at a cocktail party might evolve if the guest-list were expanded to include people of all ages, classes, races, religions, occupations, and ethnic backgrounds. No issue from infant care to incest is left untouched, yet technical jargon and highly focused ideas and discussions are banished to more specialized arenas."[41] This blurring of backstage and front stage into a middle stage is equivalent to the blurring of public and private and also homogenizes roles and places.

The power of the media to homogenize experience is important, but this power can be countered by other tendencies. For example, viewers may be active and critical in their reception of television and may see through the media, so that what is televised is not necessarily accepted as a picture of reality. Moreover, television messages are absorbed by different individuals in different contexts, thereby forming different interpretations—which contribute to increased local differences. These differences may, in turn, become the topics of television.[42] To discover how important these factors are compared to homogenization leads us back to the possible theoretical interconnections among the forces and perspectives and how these affect the circulation of meaning.

•

Television can appropriate anything and everything as material, including utopias, war memorials, and landscapes of consumption, and can circulate and transform their meanings by moving them from backstage to front stage. These scenes can also appear in, and be transformed by, advertisements, which can make programs themselves into advertisements. Television, perhaps more than any other medium, instantaneously juxtaposes place and time to create a sense of collage, montage, and pastiche, which then reverberates throughout modern culture. Television transports us to somewhere else, and it brings that place home to us. It is all the more powerful because of its apparent transparency. Despite its effects on our minds and social relations, most of us realize it is principally an image.[43]

Other features of modern culture also alter the image and meaning of place. This is especially true of commodities that, like television, juxtapose contexts. Commodities and television are also related, in that television and its advertisements circulate the images of the consumer's landscape. Television sets and their images are also parts of that landscape (they are bought and sold in places and taken to our homes). But commodities in general contain a different kind of structure than that of the television image—commodities are more basic to the geographical problem of place in the modern world. This is because commodities are not just images, they are real objects in space. Mass consumption, unlike television, can do more than alter our sense of place; it also creates real places, ones we can enter and dwell within. This is why the consumer's world is so important to the geography of modernity.

· 6 ·
A Geographical Model of Consumption

Ours is a culture of mass consumption, a consumer society as Daniel Boorstin puts it, or in Henri Lefebvre's terms, "bureaucratic society of controlled consumption."[1] We each possess different backgrounds, values, occupations, and incomes, but almost all of us have in common the fact that we are consumers living in a consumer culture. In this culture, the things we are and do often count less as common bonds than the things we consume. Mass consumption is one of the most important elements we share. Income level affects the degree to which we participate in this culture, but even the poorest among us consume and cannot escape the ubiquitous cultural images and practices of mass consumption.

Mass consumption not only produces and circulates meaning, it affects the other realms equally. It interweaves and alters forces and perspectives, and it empowers us in our daily lives to change our culture, to transform nature, and to create place. Creating ever more products to consume spurs our economy and sets in motion these relentless transformative processes. Being a consumer is thus a powerful activity. If consuming is so powerful and creative, what distinguishes it from producing?

Consumption and Production

The word *consumption* can refer to many activities: the steel plant consumes coal and iron ore; the automobile plant consumes steel; and we, as purchasers of the automobiles, consume the end products. It is the last sense of the term that is commonly used to refer to us as consumers and to our actions as consumption. But this does not mean that consumption is really the end of the production and consumption chain. Such acts of consumption create demands for new products, which in turn stimulate production. Consumer demand is as important as raw materials to this cycle. In addition, these acts of mass consuming literally produce waste and pollution, which eventually affect production and the well-being of consumers. Furthermore, the discarded articles of consumption can become raw materials in waste recycling industries, and these recycled materials eventually find their way into the consumer marketplace.

Consumption also leads to production in the sense that we use these products to create environments, contexts, or places. A family purchases mass-produced furnishings and thus acts as consumers but does so in order to create a context or environment—a living room or dining room in the home. The same family purchases clothes, and wears them to create an environment or context, perhaps to provide family members with an air of success. We also consume information, television programs, entertainment, services, and even landscapes and places. For example, an amusement park, a tourist place, a resort, a national park or wilderness area, and even a country that caters to tourists all provide experiences for which masses of people pay money. In this regard, these places are themselves consumed en masse. Even though they are not used up, they can be worn down. And when we go to such places as tourists, we are acting just like consumers of physical products and objects in that we are paying money for specific kinds of experiences. A tourist, then, is someone who intends to consume the sights and sounds of a place, and a tourist attraction is geared to produce such experiences for tourists to consume.

The consumer's world is composed of landscapes that are produced by, and for, mass consumption. They include the following: the setting or context that we create with our purchased products, which could include a home and even a neighborhood if these are mass produced; the shops, department stores, food chains, shopping malls, shopping strips, and the entire retail landscape that exist to attract consumers and that display consumer goods; the amusement parks, resorts, tourist attractions, and places that provide contexts to be mass consumed. Just as consumption forms part of a cycle with the various stages of production, so too are there landscapes that correspond to these stages. Mass-produced products are linked to these places, because consumption sets in motion the need for raw materials, which are extracted from the natural environment and require the erection of factories and urban centers over the globe and the establishment of worldwide systems of transportation and distribution.

The very act of consuming mass-produced products, then, makes us agents of production by perpetuating places and processes of production, distribution, pollution, depletion, and destruction. But the consumer's world attempts to create the impression that it has little or no connection to the production cycle and its places. It hides or disguises these extremely important geographical connections and makes it appear as though the places associated directly with consumption—the consumer's world—are all there is to this world. In other words, the consumer's world includes only the front stage of mass consumption and relegates extraction, production, distribution, waste, and pollution to a hidden backstage.

The front stage of consumption provides a context for mass consumption. We are all drawn into this consumer's world, because, whether we want to or not, we are all consumers of mass-produced goods.[2] This is most evident in the developed, or Western, world, but it is spreading as the desire for mass-produced goods spreads. With the assistance of advertising and the mass media, these goods have penetrated virtually all economies, and many developing countries aspire to the standard of living enjoyed in the West. This standard of living is difficult to separate from a consumer society.[3]

The consumer's world has been described in complex and contradictory terms. It reminds us of the world portrayed by television, but it more generally reflects the heightened contradictions of modernity. The consumer's landscape has been called democratizing, exciting, liberating, dazzling, shallow, inauthentic, disorienting, and juxtaposed. These are extraordinary descriptions, because places historically have been thought of as more or less fixed and have tended to stabilize and objectify our feelings and ideas.[4]

In premodern times, landscapes symbolized permanence, and commodities were made to last and tended to have circumscribed and shared uses and meanings. The material landscape and the commodities produced acted to stabilize and objectify the fleeting characteristics of feelings. As Hannah Arendt observes, "the things of the world have the function of stabilizing human life, and their objectivity lies in the fact that . . . men, their ever-changing nature notwithstanding, can retrieve their sameness, that is, their identity, by being related to the same chair and the same table. In other words, against the subjectivity of men stands the objectivity of the man-made world."[5] In the modern world, such objectivity is needed, especially since feeling itself is a maelstrom of contradictions, "pregnant with its contrary" and ready to "melt into air."[6] The landscapes of the consumer's world do provide this essential objectifying function, but since these places are volatile and disposable, they actually accelerate destabilization by making the contradictions of modernity more visible and real.

At the same time, this instability is tempered, because the material landscape can never change as fast as meaning. Moreover, modern places, by reifying the tensions of modernity, make them seem almost natural. The fact that these places, with all of their contradictory qualities, do change fast enough and have become so ubiquitous and enveloping draws attention, in the most visible and inescapable way, to the fluidity and instability of modern life and to the idea that perhaps this is natural and simply must be. Lefebvre echoes this role of landscape, though he emphasizes its effects on stability when he says that a consumer's landscape provides to those in society "devoted to the all-consuming transitory, and to accelerated change, the illusion of stability."[7] Precisely how the consumer's landscape works is the problem to which we now turn.

The Loom as a Model of Consumption

The key to understanding the place-building role of mass consumption lies in its complex, though unstable, integration of forces and perspectives. How does it accomplish this integration and how is it unstable?[8] Even a casual glance at mass-produced commodities shows that they embody parts of each realm. A commodity, whether a dress or an automobile, embodies social relations. It is produced and consumed under specific labor conditions and social contexts. The social history of an automobile, for example, includes the contributions of thousands of people who extract the raw materials from various parts of the world and assemble them in different places and under different conditions. From there, the automobile makes its way to retailers and finally to the consumer, who imparts to it his or her own social context.

A commodity contains elements of the natural world, because it is drawn from raw materials and becomes situated in physical space. The automobile is also an embodiment of mass and energy constructed from steel, plastic, rubber, glass, and aluminum. Once it is assembled, it is an object in physical space that helps create a context or place and that transforms the natural environment as it is used and discarded.

A commodity also contains elements from the realm of meaning, because cultures attach value or meaning to the objects they use or consume. Commodities in the modern world receive their meanings in several ways. One meaning stems from the use or function of the commodity, so that an automobile's meaning is its use as a mode of transportation, and a dress's meaning is its use as an article of clothing. Commodities also possess meaning by virtue of their price or exchange value in the marketplace, so that the price of an automobile makes it equivalent in value to anything else of the same price. And the commodity takes on the meanings the culture affixes to it: our society may suggest that certain automobiles are symbols of male virility, and that certain dresses are symbols of (male conceptions of) female sexuality.

We can mix elements from these realms, or be unaware of them, or add our own. For instance, we may add a rack to the top of our midpriced sedan and alter not only its physical characteristics but also its meaning. We may keep our car properly tuned and "car pool" to work to minimize the car's destructive effects on the environment. Whether or not we are aware of the factory and labor practices that were involved in the production and distribution of the car, our use of it will add to its social history. Simply because we are a middle-class family in a middle-class neighborhood, our vehicle becomes an important component in our daily life and in the maintenance of our social relations. And whatever the facts might be, we may have our own thoughts about the place of our car in the scheme of things.

Clearly, the commodity incorporates forces and perspectives. The connections and combinations are infinite: individuals, groups, localities, and various segments of culture all use commodities differently. But mass-produced commodities also embody an important structure that focuses these connections and helps make them public. This structure is so pervasive and accessible that, even though it can be ignored by a particular consumer, it still establishes expectations and norms that cannot be ignored by society as a whole. This element is most clearly conveyed in advertising.

Advertising

Advertisements of mass-produced commodities tell us about products. But what do they say? Sometimes they describe a commodity's attributes: for an automobile, the advertisements might include its horsepower, weight, mileage, and price. But most of the time, advertisements assign meaning to commodities. In a world of innumerable and constantly changing types of commodities, we must know what these commodities mean and do—not only, or even primarily, in the sense of their specific function, as is the case with the automobile whose purpose is to provide transportation, but in the more general symbolic functions the commodities provide. In assigning meaning to commodities and conveying this meaning to consumers, advertising can be thought of as the language of consumption. One of its messages is that commodities enable us to create our own contexts, our own worlds, however ephemeral or enduring they may be.[9] Magazine ads proclaim that H. Stern's precious gems bring "color into your life," that Alitalia "lets you live like an Italian," that AT&T "lets you reach out and touch someone," that with MasterCard International "the entire world is yours." The color of the gems, the Alitalia flight, the AT&T telephone, and the MasterCard empower us and allow us to be at the center of a milieu, an ambiance, a place.

Advertisements address virtually any theme, but their underlying theme is that commodities create contexts or places. Most advertisements embed the commodity in a context or surround it in images, which, they suggest, the commodity helps create. These contexts are places, however ephemeral. In this way, ads imply that that commodities give us the power to form similar places and to situate ourselves in the center of the world. Advertising provides a publicly shared understanding of the power of a commodity to create context and place and an idealized picture of what the context or place could or ought to be like. Consider the advertisements in the gallery in this book. One of them (figure 6) shows a truck and two people in a particular setting. The message is that if we purchase a Jimmy, not only will we be in possession of a truck (for which the price and the particular technical specifications are not mentioned in the ad) but we also can find ourselves situated in a context and partaking of experiences just like the ones portrayed in the ad. This truck will take us to a beautiful wilderness and with all the comforts of civilization. Indeed, it promises to transport us anywhere, any time, in comfort and style. This function of advertising does not require that ads be believed. It requires only that advertising be the principal public language of commodities.

Most of us are skeptical about advertising and know that we should not believe all of an advertisement's claims. And even if the claims of advertisements are believed, it is unclear how effective they are in promoting sales.[10] However, advertising is still the best way to decipher the meaning of consumption; using advertising as a text for this purpose is no more problematical than using particular versions of religious and mythical tales as texts to understand a moral system. As a representation of consumption, advertising performs a function similar to artistic realism in socialist countries, which, through the faces of smiling collective farmers surrounding their cooperatively owned tractor, portrays socialism at its best. Advertising reasserts the spirit of capitalism and becomes a mirror for its soul; it is a means by which capitalism pats itself on the back.

Advertisements as idealizations of the meaning and power of commodities might be an unintended consequence of the activities of advertising agents, whose primary motive is to promote a specific product, but this role is of paramount importance in understanding the geographical consequences of commodities. Advertising frames an idealized world; it reveals how commodities are supposed to work. In this idealized world, commodities help draw together elements from forces and perspectives.[11] Advertising points to a mechanism or structure that helps us weave together elements from all of the realms and perspectives, a mechanism that reflects the tensions embodied in mass-produced products.

This structure and its elements form the geographical quality of consumption: advertisements promise that consumption will enable us to create contexts. Each act of consumption employs one or more of these elements to construct contexts. Not all acts of consumption or advertisements need to employ all of the structure's elements. Yet due to the interdependence of these elements, most ads contain (and thus can serve as illustrations for) more than one. The Jimmy ad contains them all in a particularly well-balanced way.

We turn now to the mechanism of consumption and see how it is revealed in advertising. (The arrangement of the illustrations and their captions complement the themes in the text.)

The Loom's Structure

The mechanism of consumption can be described through the image of a loom. A loom, after all, is a structure that enables the weaver to weave threads together to create a fabric. The weaver is constrained by the materials and by the structure of the loom, but given these limits she or he is free to create whatever fabric is desired. The threads are drawn from each of the realms. The structure or mechanism of consumption is the loom and its parts. The compositions of the places created by the weaver —the consumer—are the fabric or tapestry that is woven on the loom. The metaphor of fabric and loom to describe the context or place of the consumer's world is even etymologically apposite: the word *context,* which helps describe what places are, has its roots in the Latin *contextere,* which means to weave or connect, and the fabric or tapestry is a *textile,* which draws on the same Latin stem.

Figure 2 is a representation of this loom and its relation to the forces and perspectives. The immediate problem is to show how the specific actions of mass consumption draw the elements of the realms together and transform them into place as context. This means we must specify the elements or threads that are spun of the realms, describe how the structure or loom of consumption works, and analyze the composition, tapestry, or fabric of the places created by this mechanism. We must remember, though, that these places are temporary and unstable. The places quickly unravel, to be replaced by new ones through another act of consumption. And each unraveling serves to accentuate the differences among the realms, and this in turn creates more opportunities for consumption to fill. We need, then, to explain how the place-creating mechanisms of consumption not only temporarily and provisionally draw the realms together, but also how these mechanisms and the very places they create are unstable and become instruments in transforming the character of these realms and, ultimately, in increasing their differences. I begin first with the elements or threads drawn from the realms.

Threads from the Realm of Nature

The term *nature* has many connotations.[12] It could refer to that which is real and essential or to that which is untouched by humans. It could suggest certain landscape features, such as rivers, deserts, and mountains. I use the term to refer to what the natural sciences see as fundamental components of their domain. The most basic of these components are usually expressed in abstract terms, such as *mass, force, and energy*. The natural sciences see the natural world everywhere. Because the perspective of science attempts to attain the most distant and objective position, the idea of nature becomes primarily a world external to us. We are still part of it, but mostly in the sense of being objects in this world, not in the sense of being subjects viewing this world.

My purpose here, though, is not to explore the general characteristics of this picture of nature but rather to consider the most important threads that mass consumption draws from this realm—threads that fit the purpose of consumption. Consumption uses the natural world as a resource to create products, including built environments, which are replaced by new ones and at an ever-increasing rate. We use natural resources and build, destroy, and rebuild landscapes with such rapidity that facts and events cannot be kept track of without resorting to some stable aspects of nature. For the role of consumption, the most important of these are the physical properties of space and time. They become the stable components in a world of flux.[13] Because of the focus of this book, I emphasize the role of space.

A geometrical sense of space is a useful device to refer to the underlying structure of the world "out there." Natural science concepts are often divided into those referring to substances, such as mass, energy, and the corporeality of the world, and those referring purely to space and time. These two major divisions are separated only conceptually, for space, and earth space in particular, is impossible to operationalize without corporeality, as in the relational concept.[14]

This distinction between space-time and substance penetrates the modern public view because of its relation with science and because both are grounded in the social dynamics particular to the West. These dynamics include extreme geographical mobility, which leaves us less rooted to place, private property in land, which permits parcels to be held for speculative purposes and thus to be conceived of as virtually empty space, and the enormous power to transform and move physical material from one site to another, thus uprooting things from place. Underlying all of these is an extreme territorial partitioning, which creates spatial receptacles of various sizes and porosities to contain the dynamic and mutable substances.[15] Even the academic discipline of geography, with its long-standing interest in the relation between people and environment, has developed a particular approach called spatial analysis, which simplifies the natural world by thinking of it in terms of space and spatial relations. Social forces and scientific developments thus reinforce this objective, public, geometrical conception of space as the stable element undergirding the flux of the real world.

As modern consumers, we are presented with the need for stability in a world of flux. We can use bits and pieces from anywhere in the world to create new places, which are changed with each new act of consumption. Since the material around us constantly changes, the only thing left that offers us stability is space. But here, in the consumer's world, it is not the full sweep of a lofty objective space, with its distances and directions, that comes into view. Instead it is the more fragmented picture of spatial differentiation and integration. As consumers, we integrate fragments of the world to create place, a place different from others. Such integration and differentiation provides only a partial orientation, for it does not address how this place stands with respect to others. Still, spatial segmentation and integration is a common thread from the realm of nature that is woven into the world of consumption.

Threads from the Realm of Social Relations

Social relations refer to political, economic, and social categories. Ethnicity, kinship, caste, family, worker, owner, and class are all marks of social relations. Some of these categories, such as families, are found in one form or another in virtually every society; others, such as castes, are found in only a few societies. The importance of particular social relations may change. In many premodern societies, kin and family affect most aspects of life, but kin no longer dominates modern relations, and the nuclear family, though still important, has had its influence curtailed.

In the modern period, economics plays an important role in social relations reinforced by commodities. Commodities are, among other things, economic products. Commodities cost money. They embody labor and capital in a capitalist economic system that influences virtually every facet of our lives. Capitalism's reliance on money, its facility to compartmentalize and specialize tasks, and its fostering of individualism are all well-known and supposedly related characteristics. But rather than these categories—or others, such as capital and labor, rich and poor—directly dominating the consumer's world, a distillate of these categories pervades the experience of consumption. This distillate is that we live in a world of strangers.[16] Of course, we moderns are many things other than strangers, and these other relationships influence the experience of consumption and its representation in advertising. In addition, the "strangeness" of strangers in the modern world is different from that of the premodern world and elicits a familiarity and trust not present in the premodern world.[17]

But being in a world of strangers, with the fear and even terror of not belonging, is a particularly modern facet of social relations; it pertains to any class or segment of our society and underlies practically every expression of mass consumption. Being in a world of strangers is, then, a common thread of modern social relations that commodities weave together in producing context or place.

Threads from the Realm of Meaning

The realm of meaning refers to modes of thought. Modern thought is highly specialized and segmented: it encompasses the arts, the natural sciences, religion, and myth. Some of these modes of thought are more socially prized than others. For a mode of thought to be truly modern, it must be conscious of its own structure and break with its past. In short, it must be reflexive. Such self-awareness has not always pervaded thought; we still retain many traditional conceptions and unreflexive modes. But modern thought is distinguished by reflexivity, which stems from an awareness of the human origins of thought and meaning. This awareness leads to both an acute freedom and an acute burden to create meaning: to the "loneliness of creative individualism."[18] This awareness can also lead to playful cynicism or even radical skepticism, as in the extreme self-conscious forms of thought called postmodernity.

Meaning, of course, involves the use of symbols and languages, including the arts and sciences and the spoken languages, such as English, French, and Chinese. Though these are not invented by single individuals, they are human creations, and each user contributes to their perpetuation and change. For a group then, the freedom and burden to create meaning is the freedom and burden to generate language. Even social sciences have come to use language as a model and to see behavior in dramaturgical forms. The world has always been a stage in a metaphorical sense, but now theories in social science approach behavior as though it is, literally, an act. Our performances or roles are specified by a script[19] and must be approached through the techniques of semiotics and hermeneutics. Indeed, theories of language now stress that symbols have no external referents but obtain their meanings by self-reference and connotations.[20] Preoccupation with language and our awareness of its human, not divine, origins extend the freedom and the burden to create meaning even to the definition of our own purposes in life. We are, as the saying goes, what we make of ourselves.

As we become aware of the human origins of meanings, our languages, the concepts they represent, and the lives we lead appear freer but less substantial. Like everything else human-made, they become fallible and subject to change. In this way, our freedom to create meaning lessens our convictions about the meanings we create.[21] This leads to the point where "all that is solid melts into air."

Meaning is always cultural. We are literally born into symbolic systems, which are sustained and even altered (ever so slightly) each time they are used. We learn and use these systems as members of families, communities, professions, and cultures. Modern culture produces an abundance of publicly shared symbols, so that society can orchestrate the complex and diverse behaviors required in the modern world. So successful is this culture in creating publicly shared meanings that any stranger can be plugged in, so to speak, and participate in its day-to-day signs and signals: the hellos, goodbyes, have-a-nice-days; the traffic signals; the signs of the market place and of sporting events.

These meanings are conventions, forming complex utilitarian connections in a dynamic society. They are easily shared, but they are just as easily changed; most of all, although they readily spread through society, they do not resonate throughout the culture. They are, to use Clifford Geertz's terms, "thin," not "thick."[22] Thick meanings—which are found in premodern societies and can be dissected by structural analysis—seem to be largely absent in modern societies.[23] Attempts are made to produce thick meanings, but they are difficult to sustain, especially since ritual, the primary mode for expressing thick meanings, seems out of place in the modern world. We make stabs at creating traditions and secular rituals, but these are not widely shared and often appear as "hype."[24] The issue is not whether modern people can believe intensely in symbols. About this there is no doubt. Rather, the issue concerns how much of these feelings can be deeply shared.

Resonant symbols exist at the professional, local, and familial levels, but by their very nature, these represent segmented and specialized facets of life. Thus the public realm, composed of thinly shared experiences, tends to be shallow and in flux, presenting the paradox of an objective yet impermanent world. In contrast, the private personal world, with its rich subjective content, which is not widely shared, becomes idiosyncratic. In both public and private domains, a major modern predicament is the awareness of the freedom and the burden to create meaning. This predicament provides us with both opportunities for and difficulties in communicating.

•

To sum up thus far: our conceptions of meaning have changed in modern times, and so too have social organizations and the way we use nature. These changes provide the specific content from the three realms (see figure 2). The especially modern elements from the natural realm are spatial integration and differentiation; from social relations, it is the world of strangers; and from meaning, it is the freedom and burden to create meaning. By no means do these constitute all of the significant changes. Others exist that can be called more fundamental from the perspective of one or another social theory, and these then might be used to explain or "reduce" these primary threads. A world of strangers for example can be explained along Weberian lines as a consequence of particular facets of the division of labor. Or the very same world of strangers can be seen from Marxist theory as a consequence of the role of money and commodity fetishism and, thus, as a product of capitalism. But the validity of such a reduction depends on whether or not the theories doing the reducing are true. And herein resides a problem, for as we have seen, many of these theories have important insights but are also restricted and highly problematical. The specific threads that I have identified have the virtue of speaking to many of them while not being a captive of any; they also contribute to the pattern of everyday life in a form that most people can recognize.

Even from a brief description of these elements, we can see they are dialectically interrelated. For example, geographical segmentation isolates individuals and their activities from one another and thus contributes to a world of strangers. A world of strangers, each of whom has a different domain of experience, makes meanings difficult to share. This difficulty then heightens our awareness that meaning is based on context and that we each have the freedom and the burden to create meaning. Yet this awareness and our frustrations in communicating meanings to people whose contexts differ heighten the sense that some of our meanings are public but that many are frustratingly private, and this reinforcement of the distinctions between public (objective) and private (subjective) brings us back to the problem of how to share meanings in a world of strangers.

How can these tensions be resolved? Several avenues suggest themselves. We can seek answers in basic social theories, but here again we must constantly bear in mind that these are partial and largely unvalidated, and because they emphasize one or another of the realms, a particular theory would resolve the tensions for only those who are committed to it. The fact that such theories are difficult to verify shows that they too are products of the tensions we seek to resolve; that is, the difficulties of expressing the same contexts and sharing the same meanings in a segmented world of strangers. Another avenue is the utopian one of creating a new world, or recreating some old one, in which social relations, meaning, and nature are harmoniously intertwined. But as we have seen, establishing utopias encounters the same problem that we face in validating social theories. We do not have a consensus about what and how things would work.

Of course, each of us can follow a particular theory or practice and thus try to create a realm that keeps the tensions in balance. But if these theories and practices cannot be shared and accepted by others, they will only further heighten the tensions among segmented meanings, contexts, and selves. Mass consumption is perhaps the most important means by which modern culture sidesteps these issues, at least for a while. But it, too, is problematical, because its ameliorating effects are short-lived. The patterns it weaves are, in their own way, disjointed, and they eventually unravel, so that the elements of meaning, nature, and social relations are once again separated and in tension.

The Structure of Mass Consumption: The Loom Extended

The structure of consumption, the loom, is built around a set of antitheses embodied in the relations between commodities and consumers. These antitheses are distinguishable from the threads. Yet, just as a real loom's structure must be related to the material it weaves (in that the size of the heddle holes and shuttle should be appropriate to the texture and size of the threads), there are connections between the structure of consumption and elements from the three realms—as well as a distinction between the subjective and objective perspective. The latter distinction takes us from the threads to the heddle frames. As in a real loom, these frames lift various parts of the warp, so that one or more of the threads of the three realms can pass through to be woven into place as context.

Subjective / Objective Distinction

The subjective/objective distinction pervades the three realms. Autogenic activity underpins the realm of meaning, and experience as impact is the basis of our sense of the realm of external nature. But the connections are far wider. A world of strangers and problems of shared meanings involve the distinctions between somewhere and nowhere. Indeed, an inward, private, subjective world can be fostered simply by living in a world of strangers. It makes sense for us to keep private those thoughts that are complex and ambiguous and that we doubt others share.[25] The more doubt there is that such thoughts are shared—a doubt that has foundation in a world of strangers—the more internalized these thoughts become. This expanding inner world accommodates the ambiguous and fluid component of modern life. Moreover, the fact that these inner experiences are often associated with physically distinct places thus spatially compartmentalizes the private (subjective) and the public (objective) and thereby reinforces their differences.[26]

The distinction and connection between subjective and objective is also a critical component of mass consumption.[27] Consider how a modern commodity as created and presented for sale looks from the objective side. Although in reality a typical commodity may be constructed of raw materials from specific places and may be assembled in many sites and in many stages by many different people, it normally is presented to us in a highly abstract light. In mass production, as opposed to crafts, the commodity appears full-blown, complete, and "new," often without a trace of its real origins and history. The commodity is plucked from the many places that gave it form and appears to have no place of its own. It resides in showrooms, supermarkets, warehouses, catalogues, and advertisements, which is equivalent to anywhere and everywhere. All we need is the money to purchase it, and it can be ours, instantaneously. This is precisely the message conveyed in most advertisements.

Whereas the commodity appears to have no history and resides somewhere or anywhere in space, each actual example of it must be locatable in a particular place and must possess other publicly ascertainable attributes. If it is an automobile, it is parked along with all others of its make and model in a lot or showroom and has (objectively and publicly speaking) a specific size, weight, horsepower, turning radius, fuel economy, color, and so on. But, as a mass-produced commodity presented for sale, all such cars of that make, color, and so on are interchangeable. To keep the instances abstract and general requires that the spatial-temporal system also be abstract—yet infinitely partitionable. Thus the generic nature of the product is sustained by an abstract public meaning of space.[28]

These public, objective properties of space are recognized in ads, but they are transformed to a position both closer to somewhere and yet, paradoxically, potentially anywhere—a position that lends disorientation to the consumer's world. In the Jimmy ad, the truck is in a particular place, but that place, which is not identified, stands for anywhere; one of the themes of the ad is that Jimmy is at home anywhere. Jimmy also promises to empower us with the ability to bring normally antithetical relations together—nature and culture, far and near—and to do it respectably and painlessly. Jimmy's versatility is reinforced by the total Jimmy ad campaign in which the truck appears in different settings, including a humble garage and a country club. The many real Jimmys coming off the assembly line are themselves somewhere, or anywhere, ready at a moment's notice to be ours. All we have to do is visit our local General Motors dealership.

We may be aware of the public, objective attributes of a commodity and find them attractive. We may also be interested in the commodity because we (often at the suggestion of advertising) have imparted our own meanings and values to it. Even if we do not believe that the commodity will help us attain the promises of the ad, we still have to cope with the fact that advertisements have made this commodity a symbol of such meanings and that other people are likely to interpret this commodity accordingly. Moreover, we are bound to have other feelings and values associated with the commodity, perhaps based on how it will function within our own personal context. The meanings and values that we impart to the commodity are expections, separate from the actual commodity. The instant it is purchased, these expectations will be joined to the real concrete object.[29] If the expectations are the ones suggested in the ad, then only when we purchase the commodity can the world of the ad be truly ours.

The purchase of the commodity weaves the physical, objective commodity with the subjective expectations of the consumer to form a new context. When an automobile, for example, is purchased, it is no longer an instance of a type in the abstract public domain (a vehicle with a specific size, weight, etc., in a particular location in the space-time continuum) but is also an object with meaning, creating its own place on the street or in the garage, an object or place that makes us feel excited, comfortable, or important. Both subjective and objective moments need spatial segmentation to create and support them. Objective moments need spatial segmentation in the form of an infinitely partitionable container (e.g. parking lots), able to be emptied, filled, and rearranged with an infinite variety of materials. Subjective moments need spatial segmentation to partially withdraw from the public domain in order for personal contexts (that could then be open to public view) to be created and sustained.

Threads and Heddles

The ways mass-produced commodities separate the subjective and objective components and then recombine them are represented by the large arrows in figure 2. The structure of mass consumption can be decomposed further into oppositions, illustrated in this figure by the heddle frames that form the structure of the loom. Generic/specific and apart/together are on the objective side, and we/they and self/others are on the subjective side. These oppositions are components of mass-produced commodities. Every commodity may not embody all of the oppositions equally. Collectively, the oppositions lift parts of the warp, through which threads of nature, social relations, and meaning are woven and unwoven, woven, and unwoven.

Commodities are made to appeal to many different appetites. Advertising agents, with the assistance of social scientists, have compiled lists of the most compelling needs or those most in vogue. The lists differ according to time, commodity, and advertising agency, with the result that practically nothing is omitted from the content of ads.[30] The oppositions underlie this infinite variety and help structure the diverse subject matter of ads. At this deeper level, mass-produced artifacts are driven by this structure of tensions and resolutions. These tensions and resolutions can be divided into the relatively objective and the relatively subjective, which are dialectically interrelated. They interpenetrate, and each is the mirror image of the other, which explains how the subjective and the objective are interwoven by commodities.

Consider the following subdivisions of the objective side of a commodity. Even if they are not in the final appearance of a good, the manufacture of a commodity involves the simultaneous drawing apart of some materials or places and the drawing together of others. This drawing apart and together occurs at numerous stages and scales and is true for all production, whether commercial or not. But this process is especially complex and far-reaching in the case of modern mass production. The production of an automobile draws together the raw materials of iron ore, aluminum, petrochemicals, and the labor of those who extract and process these materials from their different locations in different parts of the world. This process of drawing places of the world apart and together continues at the stages of assembly and distribution until the commodity is purchased—and even then the commodity, as it is used and moved about, may change places again and again. This (backstage) history of extraction, manufacture, and distribution is virtually obliterated when the finished commodity is presented to the public.[31]

The Jimmy ad contains no information on Jimmy's origins. Jimmy simply appears out of nowhere, full-blown. However, not all of a commodity's history is obliterated. Advertising can selectively reconstruct one or another context or stage. Tropicana claims its oranges are from Florida; Colombian coffee has its beans picked by the hands of Juan Valdez; Miller beer is "made the American way, born and brewed in the USA"; and the upholstery in a Jaguar is hand tooled. These are highly selective and often misleading reconstructions of the social and geographical history of the commodity.[32]

The power simultaneously to draw together and to draw apart is coupled with another facet of the commodity: mass-produced goods in a given product line are often similar in look, function, and price to those of other product lines, and thus they are generic in the sense that they could be interchanged (i.e., they are generic in exchange) and also generic in their use. Jimmys are virtually identical. Yet we use them slightly differently and place them in different contexts. My Jimmy is in my neighborhood, and yours is in yours. Even in this minimal sense, identical, generic products can become different from each other or specific to the place in which they are used.

As more and more commodities of a similar or generic kind crowd the market, it becomes increasingly urgent that their producers and advertisers make their brand appear to be different or specific. They might place it in unusual contexts, attractive to consumers (see figures 7 through 9, 13, and 14). As the product line becomes ever more crowded, unusual environments may not be sufficient to guarantee the specificity of the commodity, at which point ads include a personal touch, so that the commodity is directed personally to "you." While the reference seems to be to us as individuals, it applies to everyone and so shifts back to a public, objective person. The General Motors Jimmy truck, although much like Chrysler and Ford trucks and virtually identical to all other Jimmys in the General Motors line, is nevertheless—in the ad—unlike all others. It waits in the background while we are in the center; it is personal; it is "Jimmy." If we were to purchase Jimmy, even though there are tens of thousands of them, ours would be placed in a unique context and used in a way that differs from the use of all other Jimmys. Such dualism makes commodities appear generic and specific. The drawing apart/

drawing together and the generic/specific are related oppositions embedded in the objective, public side of the consumption of mass-produced commodities. Each helps specify the other.

The need to have mass-produced commodities appeal to our subjective side imbues commodities with another set of oppositions, the virtual mirror image of the two above. An appeal to our subjective side is essential, because a consumer society depends on persuading people who have not produced the commodity that they need it. Commodities can cater to any number of desires, such as sex, power and prestige, and money. The use of material objects to symbolize desires is not unique to modern culture.[33] What is unusual in the modern period is the fluidity of the association and the difficulty in finding a lasting consensus. Beneath this fluidity, though, are persistent oppositions, mirroring the objective pairs above. The tensions that portray the subjective side are concerned with self-identity. Whether or not the product claims to enhance our sex appeal, power, prestige, or income, it will claim to help us define ourselves: either to express our true selves or to change ourselves.

Self-definition is always in terms of the others who belong to particular groups, perhaps one to which we already belong (a "we" group) or one to which we wish to belong (a "they" group). One problem of self-definition is to be ourselves without losing our identity. This problem, which I term *self/others,* is the mirror of generic/specific. Just as this truck is one type among many, or one instance of thousands of the same type, so we, who are each of us one among millions of consumers, must be made to seem unique. A commodity helps us accomplish this by letting us stay one step ahead of the crowd. Yet it can also make us just like all the others who also want to be one step ahead. This opposition is common in ads. Revlon lets "the real you shine through," and J. C. Penney says "you are one of our 20,000,000 select customers to receive this offer." The extreme of this position occurs when product and self become fused and interchangeable: "Dodge is you" (see figures 12, 18, and 19).

In expressing ourselves, we can change ourselves and our context. We can be drawn from one social category to another. This capacity of the commodity to draw us from one social context to another—from being a "we" to a "they" to a "we"—is the mirror image of apart/together and is called we/they (see figures 10, 11, and 20). Although we/they is about self-expression, the opposition does not focus as much on the uniqueness of us as individuals as it does on the group to which we belong or would like to belong. Jimmy is a truck we can live with: we could be male or female, city dwellers or country dwellers. If we are city persons, Jimmy can give us an outdoor dimension; if we are country persons, it can add a city dimension. This opposition is found again and again: whether we are young or old, Pepsi makes us part of the same generation; whether we are attractive or unattractive, Chanel No. 5 will make us elegant.

These four oppositions produce tensions that can be stayed only temporarily, and each opposition can address the tensions in the others. Indeed, one dichotomy can activate another and thus set the structure in motion. (This is seen in virtually all of the ads shown in the gallery and is one of the reasons ads usually contain more than one opposition.) For instance, making generic commodities appear different can be accomplished by lending them the power to draw things apart or together, thus creating unusual and juxtaposed contexts. A commodity that places us in unusual contexts has the power to differentiate us from other consumers and help us become our true selves. At the same time, the commodity itself seems more specific, differentiable from other products.

•

This structure represents the overall mechanism in the model by which commodities are supposed to help us draw together the threads from the realms to create contexts or places. The structure by no means organizes every facet of advertising—other structures could reveal other functions. Rather, it shows how advertisements embody geographically important tensions contained in commodities. Others have attempted to go beneath the surface of ads and disclose their structure, with the intention of illuminating advertising per se and not the nature of consumption and its role in constructing places.[34] It should also be clear that the analysis extends the notion of commodity fetishism beyond its specific Marxist sense of mistakenly attributing power to objects to a more general understanding of how this power creates contexts and places.[35] Moreover, the structure developed here can be linked to social theories, if we take their partial nature into account.

The entire structure does not manifest in each ad but is characteristic of advertising in general. The likelihood that a particular commodity or its ad will contain all four oppositions simultaneously and stress each equally depends on the confluence of such factors as the idiosyncrasies of advertising executives, the nature and development of the market, and the relations among the three realms at any instant in time. The structure has particular dynamics or movements, in that some connections are more likely, given specific external factors and internal dynamics. These are difficult to separate, but I will try by considering the external first.

Historical Influences on the Structure

The structure for commodities is full-blown; all of the parts of the loom are brought into play in advertising when mass production has developed to the point where the market is flooded with mass-produced goods that are available virtually everywhere in a country, and even in much of the world. We would not expect to find this entire structure in periods before the dominance of mass production and consumption, and indeed, we would even expect the emergence of certain parts of the structure to be triggered by specific cultural forces.[36]

The following is a sketch of what can be expected to be the general steps in the development of the structure of advertisement and the major external cultural changes in the past that might have triggered them. Bear in mind that, while we are here considering the effect of cultural changes on the emergence of the structure of advertisements, the process works in both directions, so that changes in this same structure would most certainly have an effect on the culture itself—one that would accelerate the direction of social change.

Crafts

A point of departure is the historic shift from crafts to manufacture. Numerous changes are implied simply in terms of what is meant by the two words. Perhaps the most basic sense of *manufacture* is that of the production of large quantities of ready-made goods. These can be manufactured by complex machinery and techniques of mass production, but they also can be manufactured by tools or instruments not unlike those used in crafts. Moreover, manufacturing can take place under various forms of economic organization and ownership, from workers' cooperatives and factories to privately owned shops and industries. All exist at the present, but private forms of mass production are clearly dominant. Historically, the period of crafts came to a virtual end for some economic sectors by the late eighteenth century but persisted as the dominant form for other sectors well into the nineteenth century (by which time, a move to resurrect lost crafts began, though it eventually became commercialized).[37] More precise dates can be identified, sector by sector, for different regions and countries, but a general index of the development of this shift can be seen in the transformation of the geographical landscape.

The emergence of mass manufacturing has left its mark geographically, especially in the city. Urbanization and industrialization, though distinct processes, are nonetheless closely related, especially in the economically developed world.[38] Among the factors connecting industrialization and geographical organization are the increases in population that attend industrialization, the increased separation of work and home, and the development of warehousing, wholesaling, and retailing as geographically distinct from manufacturing. Before the rise of industry, the typical city was smaller, its population was smaller, and it had far less internal differentiation. It was a pedestrian city of wards, districts, or neighborhoods, each providing a mix of social classes and economic functions, and not a highly spatially partitioned world of strangers.

If we take the American city as an example, it is clear from David Ward's work that "until the mid-nineteenth century, apart from the small exclusive residential quarters of the rich, the functional specialization of urban land uses was only weakly developed. Most industrial and commercial activities generally were conducted on the premises of the producer or merchant, and local purchases or services were obtained on a custom basis directly from the producer."[39] In the era before mass production and mass consumption, the producer was in close proximity to the consumer and often made products on demand. Advertisements were expected to exhibit virtually none of the elements of the full-blown structure identified for the contemporary period (see figure 3). Instead, advertising took the form of signs, handbills, and newspaper displays. These began to appear in the late eighteenth century.[40] Advertising was not yet a trade (the profession began in the 1870s), and the craftsman (usually a male) himself made his own advertisements or hired an artist to help with the design. Advertisements did not focus on the commodity, for these were often custom made, but rather proclaimed the skills of the craftsman. Ads often included sketches, but these were meant to stand as emblems of the crafts rather than as representations of specific objects that could be purchased ready-made.

Standardized Production

Gradually, standardized production superseded crafts. The process took place in degrees and at different times for different sectors, with the post–Civil War period an important watershed for many industries in North America.[41] Standardization of production assured that each instance of a commodity was of a standard quality. This factor in itself often made the purchase of a ready-made commodity more appealing than a craft-produced one, whose quality could vary considerably from others made even by the same craftsman. It is to be expected, then, that the attributes and qualities of the commodity itself, rather than its producer, became the object advertised. For the most part, the ad presented an object alone, without context. Often its characteristics were described, its price mentioned, and a sketch of the product included (see figure 4). Context was not required, because most early ready-made or machine-made products duplicated those that were produced by craftsmen. Thus they fit into well-established webs of use and social context, and there was little need to employ a novel situation to set one product off from another. The novelty of mass-produced products, their competitive prices, and the standardization of quality were attractive enough selling points.

Mass Markets

Once the mass market for particular commodities became crowded, which began for a wide range of goods in the late nineteenth and early twentieth centuries, or once entirely new commodities were introduced, it was important to have them stand out, so to speak, and their purpose defined. Placing commodities in contexts then became important.[42] Ads sold contexts as well as commodities—the sizzle and not just the steak. Still, the turn of the century was not as much a world of strangers as a world of class and occupational groups, and so the contexts in ads were often specific, representing real places or precise social situations.[43] Sometimes people were portrayed, but there was little overt appeal to the commodity's potential in defining and expressing self. Undeniably, the consumer had expectations about the commodity, but the ad did not focus on the subjective. Until the early twentieth century, despite considerable social mobility, most people still derived identity from their positions in specific classes and occupations.[44]

Mass Production and Social Mobility

After World War I, social mobility increased and so did the numbers and types of mass-produced commodities. Increased social mobility means more fluid and often impersonal relations, and this kind of world, where most people know each other casually or are simply strangers, needs to draw our attention to how products can place us within a group or assist us in finding our true selves. Ads had to appeal to tensions of we/they and self/others, in addition to generic/specific and apart/together. These tensions would remain important because the similar commodities flooding the market indicated the need for ads to differentiate among virtually identical commodities so that the generic could appear specific, and new commodities could be separated from some things and combined with others. When all four oppositions are present, the tensions in each activate the others. This is, of course, the kind of advertising that now exists, and its range indicates all four tensions operating.

•

This historical sketch suggests that the objective side of the opposition precedes the subjective side and that the generic/specific precedes apart/together—which precedes we/they, which precedes self/others. This sequence points to potential connections. The particulars of the sequence can be altered by innumerable considerations, such as the fact that commodities differ in terms of the periods of their introduction, the conditions of the market, competition, and the social contexts of the times. Perfumes, razor blades, and patent medicines were set in unusual contexts or appealed to personal fulfillment early on, while furniture ads began much later to use such methods. Other factors affect the details, relations, and timing of ads, not least of which are the technology, medium, and artistic and graphic styles of advertising and the ideologies of advertising agencies, manufacturers, and the competition.

Internal Dynamics

External relations put oppositions into place, but once activated, these oppositions create a dynamic of their own, which affects the balance or quality of the oppositions themselves, the character of the threads, and even the external realms of nature, meaning, and social relations. The internal dynamics of oppositions are adduced from the structure's own logic, but they can be triggered by the broader cultural setting.

Internal motion, on the objective side of the structure, can be set off by some external factor working on one of the oppositions. For example, in a saturated market, the need to differentiate among commodities is heightened. This need can lead advertising to set a commodity off from its competition by making it appear to be able to draw more disparate things together to make it seem that it can create more unusual contexts. As the markets for these commodities begin to include people with ever more diverse backgrounds, the contexts or places portrayed in advertising become more abstract and generic. Particular scenes, whether in the country, the city, a building, or a room, are juxtaposed with other scenes in fantastic pastiches. With each attempt at uprooting and juxtaposing contexts, the original setting loses more of its depth and complexity (see figures 24 and 25). This loss can be thought of as a cheapening of symbol and context, which was feared to be one of the effects of mass popular culture even in the nineteenth century.[45] The loss can also provoke general observations about the inauthenticity of meaning and place in the modern world. These changes in context are accelerated by advertising's own technology, which includes vast files of places that can be combined and edited to create hybrid and juxtaposed contexts.

Parallel movement can be expected on the subjective side. As our sense of self becomes appealed to more and more, and as we are continuously presented with the power of the product to help distinguish ourselves from others, ads will increasingly segment the self. Ads that claim they can place us in or out of context fragment meaning and increase indirection.

The consequence of these internal changes is to throw each opposition into imbalance by accentuating one side over the other; in terms of the loom, each heddle would move in one direction more often than in the other. This imbalance affects the quality of the oppositions. Apart/together could lead to increased juxtaposition of place; generic/specific could lead to increased generality of place; self/others could lead to an increased fragmentation of self; and self in/out of context could lead to vagueness between we and they. These changes in the quality of the oppositions presuppose that the entire structure is present, and these changes do not progress according to any particular sequence. There is no reason to expect that increased generality will precede or follow increased reliance on self-identification with the commodity or that increased juxtaposition of place will precede or follow a weakening of the distinction between we and they. The only implication is that the qualities will shift in this general direction despite temporary reversals from shifts in meaning and the increased targeting of advertisements to particular groups.

We would expect there to be a historical development in the structure. Once all of its parts are in place, the quality of advertising will shift toward an emphasis on generality and juxtaposition of contexts, greater fragmentation of self, and a greater vagueness between we and they. These relations can be examined over groups of products or fields or for the life of an individual commodity. Most historical studies of advertising were not designed to test either the internal or external relations, per se, but they do disclose trends that are much like the ones hypothesized here.[46] Precise tests would be difficult to develop because of the problems of isolating tensions and even products. A single commodity can accumulate contradictory meanings as its advertisements change. A commodity can also rely on other ads to provide context. If ads for similar commodities fantastically juxtapose or rearrange things and claim to do it all for us, an advertiser could omit this sort of appeal and present the commodity in stark and "realistic" terms (see figure 15; compare with figure 16). We should not forget that other considerations intrude, including the individual advertising styles of the particular media, advertising technology, and the workings of external cultural forces.

Advertising is not an end in itself; it exists to sell commodities. From the producer's view, this is its single most important effect. But from the geographical perspective, the effects of advertising go beyond the act of consumption. Advertising and its dynamics impart meanings to products and help make the commodity a mechanism or structure drawing forces and perspectives together. The commodity's function is inextricably linked to this drawing together of forces and perspectives. Even if we wish to resist its function by using the commodity differently, we still have to take this structure into account. This dynamic structure of advertising does more than situate the commodities within the realms; it also affects these very same realms. I consider next the way advertising alters the material it draws upon.

External Dynamics

How do the dynamics of consumption as portrayed in advertising affect the particular material or threads from the three realms—and even our view of the realms themselves? Remember that tensions exist among these threads even before they are woven by commodities. For instance, geographical segmentation contributes to a world of strangers, and a world of strangers makes meanings difficult to share. This difficulty heightens our awareness that we have the freedom and the burden to create meaning, and yet it further reinforces distinctions between subjective and objective contexts, which brings us back to the problem of shared meanings in a world of strangers. The weaving of these threads by commodities accentuates these paradoxes. I consider first how the threads are affected and then reconsider their dynamics and the impact these changes have on the realms, on the perspectives, and on culture.

The Realm of Nature

Consumption transforms nature and yet disguises this transformation by creating a contextless nature. The term *nature* signifies the objects and forces in the domain of the natural sciences: electrical, nuclear, and mechanical forces; rocks, animals, vegetation, and space. All of these have been drawn upon to create commodities, and each by itself, or combined with others, can be a commodity. The continuous and capacious transformation and commoditization of forces and objects place the elements of nature in a state of flux and impermanence. One of the few elements that remain stable is the abstract sense of space. This is why an infinitely partitionable spatial continuum that can be segmented and integrated is the thread I identify from the realm of nature. Commodities, though, provide their own twist, even to this continuum.

Segmentation and integration in the consumer's world occur through the commodity's power to draw things apart and together, so that we can use it to create place as context. An advertisement's representation of the commodity and its surroundings describes the material setting in a single instant. Though ads cannot represent things without distortion, for the sake of argument let us assume that an ad's representation is accurate. Still, the advertisement frames this place with no reference to its location or its past and with no connection to other places either in the real landscape or in the content of the medium where the ad appears. Ads present a segmented world that makes sense only in the frame of the ad. There is no true backstage in this world. Space is annihilated. This leads to a dissociation and lack of orientation with whatever preceded it or is adjacent to it in the "real" world and even in the context of the medium where the ad appears.

Place or context in ads are strangely placeless, which makes even the real places portrayed in ads lose their meanings. This is increasingly the case when an ad is a collage of real places. A realistic looking cityscape may be a collage of buildings from dozens of other cities (see figure 24). A famous geographical site such as the Statue of Liberty may be placed in another part of the world (see figure 25). Such juxtaposition suggests that the material world is in flux and that virtually anything can be moved and transformed by the power of consumption, so that nothing is necessarily of a place. The liberty that advertising takes with geography and history, by uprooting and juxtaposing images of place, oddly enough makes it all the more important for the world beyond the consumer's world to embrace an abstract geometrical continuum as the public, objective meaning of space, which pushes this space even farther away.

Advertising not only uproots and juxtaposes places, but also alters the meaning of nature itself. The natural sciences conceive of nature as simply out there, infinite, and independent of human action, forming the wild, untamed world. But advertising circumscribes and tames this nature by making it into a commodity. To borrow Claude Lévi-Strauss's terms, "raw" nature becomes "*cooked*,"[47] or at least half-baked. This is also true of space itself, which can be conquered, annihilated, tamed, or manufactured. "Do you yearn for more space?" an ad reads. Circumscribing discrete places on the earth and calling them nature or wilderness and presenting them as accessible scenes and landscapes that can be viewed and experienced tames an infinite and independent nature and places it under our control (figure 23). This leads to a paradox, for a natural world packaged as places to view and experience is no longer untamed and untouched—and yet it is attractive precisely because advertising makes it appear as though it is untamed and untouched. Moreover, packaging nature diminishes its scope. It can be absorbed by anyone in a finite amount of time. These effects raise the ultimate geographical irony: Are these natural places in fact inauthentic?

Advertising also affects how we conceive of nature by projecting its tensions onto the subject of the ad. Thus when a natural place is the subject of advertising, the portrayal of that place will go through the same tensions as in ads for manufactured commodities, but with a slight inversion. In the case of places, which are by definition different from one another or unique in location (whereas manufactured commodities are similar to one another, or generic), ads first show how these places are like other kinds of places and readily accessible (i.e., they contain hotels, roads, and running water) and so the ads portray them in a general light. Only later do the ads present these places as unique or specific, though by now, after the homogenizing effects of tourism, in a somewhat fabricated sense (with re-creations of specific cultural features, such as native dances and artifacts, that have been transformed). Even later, the place becomes essential for our self-identification and self-expression (see figure 18, 19, and 20).

The Realm of Social Relations

The places or contexts that commodities create are social as well as material. Consumption purports to dispel the dread of being in a world of strangers.[48] Ads tell us what to expect, what is acceptable and unacceptable, what we need to do in order to belong. They are primary vehicles for producing and transmitting cultural symbols. May we wear a tie and jacket or an evening gown while driving a truck? May we drive a truck to a country club? Is it proper to wear designer jeans to a restaurant that has real linen tablecloths? What kind of alcohol should we expect to be served by our best friend?

Commodities themselves, in other words, help build social relations. This empowering capacity of commodities is one meaning of the terms *consumer society* and *consumer culture*. Possessing the correct product can be an entree into a social situation. By providing this kind of access to virtually anyone with the money to purchase commodities, social relations become democratized in that background, class, ethnicity, and race become less important than the possession of the correct clothes, car, or credit card. This does not mean that racial, ethnic, and other differences are not portrayed in advertising. They are, but not to inform us about social reality or to raise our consciousness about injustice, but to set one commodity off against another: this kind of suit makes us members of the upper class, this kind of truck makes us outdoorsmen, and Pepsi puts us all on an equal footing as part of the same generation.

The social contexts of ads are almost always pleasant. Products are rarely connected to unconventional and uncomfortable social relations.

Unlike nature, which advertising portrays with extraordinary freedom, social relations are presented conservatively. It simply does not make sense to claim that a product helps create awkward social situations, that it makes us indistinct, that it prevents us from belonging—unless the advertisers are claiming in a negative ad campaign that these are the consequences of consuming the competitor's products. And it does not make sense for advertising to set the commodity in a social context that is not acceptable to the targeted market.

Advertising draws on our dread of being in a world of strangers. By making us consumers, it allows us to define our own social group, our own selves. Even when the general public accepts these social cues (and it is remarkable how hundreds of millions of people can all at once latch onto several among thousands of competing cues and agree to call them the culturally accepted ones of the moment, as in the meteoric rise of the Cabbage Patch Doll), these cues do not solve the problem of being in a world of strangers. The introduction of new products and the repackaging of old ones makes for a constantly changing consensus—or no consensus.

Each time the commodity draws into a context, it leads to a different self. A commodity can represent the entire self or its tiniest part. Imagine all of this year's ads for cosmetics. Suppose that each advertisement contains a picture of a model using the product in a particular setting. Suppose that all of these images can be arranged around a picture of a real consumer, with lines connecting parts of her body—hair, scalp, eyelids, face, lips—to the ads that correspond to those parts. If we follow these lines outward from the consumer's body to the advertisement of the product and then to its context in the ad, the self is fragmented into an infinite number of parts; each part, through the purchase of a corresponding product, transforms the self. Each context, then, defines a different self, segmenting us until there is no consistent self left. We are defined and redefined in context after context.[49] In G. Debord's terms, self (and social relations) move from being, to having, to appearing.[50]

The Realm of Meaning

The realm of meaning refers to the operations of the mind, and the most consistent manifestations of thought are symbolic systems or languages. Advertising is the language of consumption; the thread it draws most from the realm of meaning is that of the freedom and the burden to create meaning. Advertising is a more hybrid language than most. It borrows material and forms of expression from anywhere and everywhere, but its primary rhetorical device is indirection.[51] Commodities promise to create contexts and places, to link us to groups, and to communicate meanings. While this is undeniably their message, advertisements do not say this exactly and through declarative sentences. Even when they use ordinary language (and even the specialized language of science, as when an advertisement's claim that an electronic shaver uses advanced technology is supported by placing the shaver next to an incomprehensible picture of a printed circuit),[52] they do so through innuendo and indirection. Simply drawing attention to a commodity in a place suggests that the commodity creates the place.

In other modes of discourse, indirection can consist of a single contrast, as with most metaphors and similes. But advertising is different again, in that each statement contains at least one missing element, the consumer.[53] Each statement is incomplete unless we, the consumers, act—unless we purchase the product. This principle of incompleteness also sets advertising apart from art. Art is everywhere in advertising. Indeed, more artists find employment in commercial arts than in any other sector of the economy, and much advertisement has aesthetic qualities. But unlike a successful work of art, which does not have to inspire us to act, a successful advertisement must lead us to consume.

As a symbolic system, consumption is perhaps most like conventional magic and ritual.[54] Both advertising and magic or ritual impute fantastic power to objects. Occasionally, an advertisement portrays its products as literally possessing magical properties (see figures 22, 26, and 27). Both ads and magic claim we can tap these fantastic powers if we undertake the actions prescribed. But unlike ritual, which promises specific results in a public arena of conjoint actions (a situation that is nurtured by small and traditional societies that reinforce shared conceptions of symbols), advertising most often makes its claims by indirection, to us in our homes through the privacy of the radio, the television, and the magazine. It provides a seed for thought, and we have the freedom to use the product in unprescribed ways.

Unlike magic or ritual, whose forms and effects are fixed by tradition, those of advertising are open-ended and constantly changing. At one time a product promises us this, at another time it promises us that—and the competition claims that it won't provide us with anything at all. This mutable quality of the symbols of ads, as opposed to conventional magic or ritual, is based on the differences in the conceptualizations of symbols in the two languages. A symbol, whether a word, phrase, thing, or gesture, stands for a something else, which need not be physically present.

In magic or ritual, a symbol is so closely linked conceptually to the absent thing it represents that it can literally stand for, or be, that thing. Thus in voodoo ritual, a doll that stands for an individual can literally become that individual. Or in more conventional religions, the taking of the wafer and the wine in the ritual of communion makes these objects not just symbolize Christ's flesh and blood, but become them. Becoming these objects, or being conflated with them, makes such symbols powerful. We can affect the referent by affecting the symbol; conversely, the symbol can channel and capture the referent and its power. This conflation of symbol and referent can be expected most when we perceive a resemblance between the symbol and its referents—perhaps a similarity in appearance or shape—and when the community is close-knit, so that such perceptions are reinforced. This conflation makes the symbols in magic or ritual appear to be natural, and this appearance of naturalness makes the meaning of such symbols nonnegotiable.[55]

Geographical space and place are often the principal symbols of power in conventional magic or ritual. Topographical features such as unusual landforms, high mountains, springs, and caves and the directions north, east, south, west, up, and down stand for the forces of good and evil, birth and death, fertility and infertility, and so on. As symbols of such forces, they become the forces themselves. Thus cardinal directions can have power, and so too can particular places. This means that, to tap these powers, we must be at a particular sacred place and facing in a certain sacred direction. In ritual or magic, there is then a necessary, not a contingent, connection between the events undertaken and their locations and orientation. Such is not the case in advertising. Even though its empowerment of commodities is fantastic, it does not have this unalterable conflation of symbol and referent. Its symbolic meanings are more fluid and even playful, and space and place do not possess magical import.

Advertising's rhetoric of indirection, its capacity for change, its brevity, its ability to make a pattern when none exists all help to make advertising applicable to anything, even to items that are not strictly of the marketplace, such as conservation and gay liberation. But once advertising embraces these issues, it transforms their meanings. They become commodities; they become segmented and abstracted from context. Advertising creates its own skepticism. We may know that the claims of ads are not really true. We may know that objects do not really possess these powers. We may know that they will disappoint us. Public cynicism is revealed in the ads themselves, as when an ad begins, "This is not just another ad." This reflexivity, which sets advertising apart yet again from magic or ritual and other languages, stems in part from a doubt about the veracity of its claims and about the virtue of its purpose. But it also reflects the fact that we are conscious that ads are made by men and women. All symbol systems are constructed, but advertising is nakedly so. Its structural oppositions make its human origins clear. A recognition that a major language of our culture is human-made heightens our awareness that a modern predicament is the freedom and burden to create meanings.

Perspectives

Meanings and perspectives are related; advertising affects the latter by altering the distinction between subjective and objective. It purports to unite subjective feelings and objective things. But the dialectics of consumption require that new desires be created and satisfied by commodities. In the long run, this aggravates tensions between subjective and objective. Real commodities, as objects in a public, objective world, must be produced, distributed, and purchased in as many places and as quickly and efficiently as possible. This means that public, objective space becomes more and more a geometrical continuum, forming the backdrop to the related and dynamic qualities of production and consumption. We each possess territorially segmented personal worlds, which the commodity helps to fill, but when we bring these commodities home to our personal worlds to give us something of the contexts they promise, their meanings become altered. This makes them more difficult to share, which then heightens the difference between subjective and objective. The distinction between subjective and objective is made more complex by the fact that products multiply the number of selves a subject can have or, what amounts to the same thing, fragments the subject into innumerable parts. Thus, in attempting to join subject and object, commodities isolate, multiply, and fragment the subject.

•

Consumption undoes contexts to create contexts, undoes social relations to create social relations, and undoes meaning to create meaning. The segmentation of context and the isolation of self make the complex and personal experiences and meanings of modern life difficult to share. We attempt to bridge the realms through social theories and through actions that reflect one or another perspective, as when the confidence of social science leads to the brave new world of glittering skyscrapers and planned cities; or we recoil from such efforts to dominate by withdrawing from society and creating rural utopias. These theoretical and practical efforts, though, become highly specialized and often do not address everyday life. The more segmented that meanings become, the more aware we are that we create them.

We know that advertising is created in places like Madison Avenue, that advertising is hype, and that its purpose is to sell. This awareness and the unfulfilled expectations that purchasing commodities can bring present a constant threat to the credibility of consumption and threaten to unravel its fabric. If the fabric does unravel, we return to our starting point: the general problem of synthesizing the three realms. The paths back to the three are seen in the ways we resist consumption. We can disclose the negative effects of consumption on nature; we can reveal the social exploitation of the processes of production and distribution; and we can identify the emptiness of its meaning. Graffiti, as an act of resistance, can even point these out in the ads themselves (see figure 30). To support any of these routes, we must once again enter the three realms and choose from among the incomplete and largely speculative theories in which one mode assumes a greater explanatory force than another.[56] Or we can again turn away from theory and become immersed in the routines of day-to-day life, which depend on mass-produced commodities to create contexts and places and which temporarily stay the tensions of modern life. In other words, we can become consumers.

Advertising is not an isolated event. It is one of modern culture's devices to give things meaning. Advertising is a blueprint of how commodities create places. I turn now to an analysis of this place-creating process as it actually occurs in the geographical landscape. I show that the same structure revealed in advertising is contained in, and reproduced through, concrete geographical places. Places of consumption, in a very real sense, are three-dimensional advertisements that not only sell commodities but sell themselves. These are the places we encounter in daily life, and because they are real and concrete, they have even more weight than forms of representation like advertising.

· 7 ·
Places of Consumption

People create places and are affected by them. Each type of place contains different mixes of forces and perspectives, but the weave of places of consumption is especially significant because these places are so prominent a part of modern everyday life and have so great a bearing on the structure of modernity. Much of our awareness of the modern, and now the post-modern, stems from the manifestation and intensification of these phenomena in places of consumption. Modernity is lived and felt through these places, whose landscapes look and feel different from yesterday's.

Places of consumption interweave and alter elements of meaning, nature, social relations, and agency. They do the same for perspectives. They stand in an intermediate position between the public and its objective, geometric, but often alienating, space—expressed in the landscapes of straight-lined railways, highways, skyscrapers, and represented to the public in the grids of maps—and the private, and its subjective, idiosyncratic sense of place. The power of consumption to forge these disparate elements into places that share the same qualities is most evident in the fact that places appearing to contain different mixes of private and public, or of nature and culture, will in fact turn out to have much the same structure when reworked at the hands of consumption. This power of consumption to forge a common structure is examined in the contexts of department stores, retail chains, shopping malls, homes, workplaces, and recreation places away from home—in tourism and travel, and in the reality of Disney World, and museums.

Stores and Shopping Malls as Commodities

Stores contain the commodities described in advertising. The store's environment must be attractive for these commodities to sell. This means that the store acts as an advertisement for the commodities, displaying them in a way that makes them as attractive as they were in the media ads. Hence the attention to store layout, lighting, window displays, background music, and a courteous and efficient staff. As a stage setting for the display of commodities, the store may use enlargements of visual ads as backdrops for the commodities, and it may display them in such a way as to form actual contexts on the store floor. This leads the stores themselves to possess the tensions that are contained in the commodities.

The sense of a store as a context for consumption is accelerated and extended to a larger environment when stores take advantage of their proximity to other stores in business districts and especially in shopping malls. This proximity increases the chances that we will window-shop or wander into the store to browse, as we might thumb through magazines and pause at the ads and eventually be prompted to buy—perhaps not now, but at some subsequent visit. Just as ads do not, for the most part, represent commodities that we "must" use right now but rather create needs and desires for commodities that we may never have thought of or desired before, so stores and complexes of stores become contexts for increasing our appetites. We may be shopping for a particular commodity but in the setting of the store we are titillated by the displays of other commodities, which we might come back to buy.

Commodities, stores, and clusters of stores become landscapes that advertise both particular goods and consumption, in general. Because sales are enhanced by window-shopping, browsing, and exposure to an environment that titillates the senses, the landscapes not only contain commodities that can be consumed, but the landscape itself is being consumed. As both advertisements for and containers of commodities, the landscapes embody the tensions of these commodities. This is indeed the current state of affairs in much of our landscape, but this merging of commodities in place with places that can be consumed has a history, just as advertising and its tensions has a history.

Department Stores

Stores always contained commodities, but until the late nineteenth century they were called shops and mostly stocked a narrow range of merchandise: shoes in a cobbler's shop, baked goods in a bakery, drugs in a pharmacy, meat in a butcher's shop, hats in a milliner's shop, cloth in a tailor's shop, and so on. The shops often sold things made on the premises, often to the specifications of the client. Prices were not preset but depended on the customer's needs, the craftsman's skills, and negotiations between the customer and the shopkeeper. People patronized shops in their neighborhood, because transportation in many of the older cities was difficult. Even a journey by foot was difficult due to twisting lanes, cul-de-sacs, and clogged main thoroughfares.[1] In addition to proximity, people selected shops based on their reputation for price and quality. The shops did not have large displays of wares. Indeed, entering a store meant that the customer intended to order or purchase, or at least seriously haggle over, a commodity. Browsing was more a part of open-air markets and bazaars.

The advent of the department store in the second half of the nineteenth century made a striking change in the way commodities were sold. Among the most significant of these new types of stores was Bon Marché, established by Aristide Boucicault in Paris in 1852 (see figures 31 and 32). The department store was novel in several respects: price markup was small, volume was large (though Bon Marché first specialized in dry goods and fashions), and commodities were sold at a fixed price that was plainly marked.[2] These innovations were soon found in other major cities: Lord and Taylor, Arnold Constable, and Macy's in New York City; Potter Palmer and Marshall Field's in Chicago; John Wanamaker in Philadelphia; and Burt's and Whitely's in London. High volume, low markup, and a plainly marked price, all of which lead to a large turnover, would not have been feasible if manufacturers were not producing large quantities of a wide range of standardized commodities[3] and if there were not reliable and efficient transportation systems: railroads, wide and straighter urban streets, horse-drawn buses, and electric streetcars, all of which could be used by manufacturers, distributors, and the consumers who patronized these new department stores.

Department stores provided customers with free entry, with no obligation to buy; thus they had to sell large quantities of goods to people who may not have intended to purchase anything. Department stores attempted to increase consumer appetities by serving as advertisements for the commodities, arranging them in such a way as to cause a large turnover. Stores became spectacles, "endow[ing] the goods, by association, with an interest the merchandise might [otherwise] intrinsically lack."[4] Decorators "'transfigured' and 'transposed' the stores as well as the goods into 'pictures' to impress the customers."[5] These pictures gave the sense of department stores as magical cornucopias or paradises; they offered everything to everyone without effort.

Just as the foods pouring from the mouth of a cornucopia are in no particular order, as though to emphasize the indifference to order that comes with opulence, so department stores use juxtaposition as a principal strategy. "The first means the retailers used," according to Richard

Sennett, "was that of unexpected juxtaposition. . . . Rather than one hundred pots of the same size by the same manufacturer on the floor, there would be one example only, put next to a pot of a different shape."[6] Émile Zola points out that "the strength of the department stores is increased tenfold by the accumulation of merchandise of different sorts, which all support each other and push each other forward."[7] His novel about department stores, *Au Bonheur des Dames,* was an attempt "to create the poetry of modern activity." The same claim is made by G. D'Avenel: "It seems . . . that the most dissimilar objects lend each other mutual support when they are placed next to one another." Why should this be? Sennett says that "the use character of the object was temporarily suspended. It became 'stimulating,' one wanted to buy it, because it became temporarily an unexpected thing; it became strange."[8] Entering Bon Marché, people experienced something that was part opera, part theater, and part museum.

> Bargain counters outside entryways produced a crush at the doors that attracted still larger crowds, thus creating for all the sensation of a happening without and within. . . . Everywhere merchandise formed a decorative motif conveying an exceptional quality to the goods themselves. Silks cascaded from the walls of the silk gallery, ribbons were strung above the hall of ribbons, umbrellas were draped full blown in a parade of hues and designs. Oriental rugs, rich and textural, hung from balconies for the spectators below.[9]

Juxtaposition, or the "chaotic-exotic," in Grant MacGracken's term became the general practice of department store displays.[10] Sennett claims that the excitement produced

> by jumbling dissimilar objects together the retail owners reinforced by the continual search for exotic "nouveautés" to put on sale in the midst of the most prosaic wares. Strange goods . . . are useful not only as articles of trade in themselves. They accustom the buyer to the notion that he will find in the store what he did not expect and thus be willing to leave the store with merchandise he did not enter in search of. Volume . . . was achieved in the retail trade through an act of disorientation: the stimulation to buy resulted from the temporary well of strangeness, of mystification that the objects acquired.[11]

Disorientation and high volume had yet another effect. The hustle and bustle of consumers making numerous purchases made it appear as though the products would soon be gone. This illusion of scarcity (it was an illusion, because the products were, after all, mass produced) further stimulated the consumer to purchase a commodity now, rather than wait until it might be too late.[12] This disorientation was so intense that it may even have caused shoplifting. One shoplifter explained the disorienting effects of Bon Marché:

> Once plunged into the sensuous atmosphere . . . I felt myself overcome little by little by a disorder that can only be compared to that of drunkenness, with the dizziness and excitation that are peculiar to it. I saw things as if through a cloud, everything stimulated my desire and assumed, for me, an extraordinary attraction. I felt myself swept along towards them and I grabbed hold of things without any outside and superior consideration intervening to hold me back. Moreover I took things at random, useless and worthless articles as well as useful and expensive articles. It was like a monomania of possession.[13]

Department stores were becoming advertising showcases. As Jerome Kerber, a twentieth-century store decorator put it, to draw attention to the commodities, the point was to "eliminate the store."[14] Not only was Bon Marché a stage setting or advertisement for goods, it soon became a pioneer of conventional advertising through newspapers and even postcards and calendars, which presented the store as a fanciful and magical place—a dreamworld.[15] The store became such an attraction in itself that it was included in sight-seeing tours of Paris and was prominently displayed on city maps. The Bon Marché soon added luster through such adjunct services as a reading room, an art gallery, and house concerts. Soon the rhythms of the year, the seasons and holidays, were adopted by the store as sales motifs.[16] This made it possible for the store to be not only a source of merchandise but also a source of "culture," whose luster illuminated all of its merchandise. By purchasing a commodity from Bon Marché, people participated in and purchased this culture.[17]

Department stores quickly spread to all the major cities of Europe and North America, and with them came their power to link commodities and contexts, not simply through the force of suggestion that the consumer remembered from reading ads but in a tangible form in the store itself. In this respect, the stores became living ads, and the consumer who entered their premises was actually directly experiencing the effects of the power of commodities to create contexts.

The department store drew on the achievements of the times. Developments in manufacturing and transportation were indispensable ingredients in the success of department stores. But the stores themselves contributed to the technology of business. They were pioneers in management and organization; each one employed hundreds of salespersons and hundreds more people behind the scenes. They were major forces in the development of credit. Their backstages of warehousing, packaging, advertising, and planning were as important and novel as the front stage of the display and sale of commodities. They introduced new management practices and forged new labor relations. The sexual divisions of labor—women as salespersons and men as managers—were to have far-reaching effects on labor practices in the rest of retailing.[18] In some cases, the stores were paternalistic, providing medical assistance, pensions, and even room and board for some of their employees, in exchange for loyalty and obedience to store rules and a sense of decorum.

Department stores also played an important role in the social and psychological transformations of the late nineteenth and early twentieth centuries. The increase in social mobility and in the scale and complexity of the economy threatened older, smaller, and more local forms of social organization. Taking their place were mass society, mass culture, and a world of strangers. This public, objective world made it difficult for people to feel at home. It was cold, analytical, and remote from their personal, private worlds, which became separated from each other and awkward to share. Mass-produced commodities and mass consumption provided a bridge between the private (subjective) and the public (objective). This bridge enabled consumers to create contexts that communicated their feelings and fantasies in a public language.

The department store became the geographical locus of this bridge. It was also a "democratic" institution in terms of economic classes (though not races), allowing the rich and the poor to enter and participate. Yet it created its own contradictions; one of them was that, although it enabled consumers, it did so by making them more passive. This passivity stands out when we compare the skill and effort involved in haggling in markets and shops with the effortlessness of purchasing products in department stores with fixed prices, willing salespersons, and abundant displays that inform us how this or that product would help to recast our lives.[19]

Department stores soon were to extend their influences geographically by establishing mail-order catalogues and branch stores. As early as the 1870s, Bon Marché ran an extremely profitable mail-order system, employing hundreds of clerks.[20] Mail-order catalogues could, however, be the sole purpose of a business, as was the case with Montgomery Ward, and Sears and Roebuck. The Sears and Roebuck catalogue, often referred to as the Great American Wish Book, soon became a pioneer in displaying in a catalogue an entire range of items, from clothes to furniture, kitchen utensils, horse and buggy accoutrements, and later automobile accessories.[21] If the business was solely a mail-order business, the catalogue, as a compendium of advertisements, had to substitute for store displays. But in most cases, a business had both a store and a catalogue. Store layouts and layouts portrayed in the catalogue often coincided. These images of contexts were then duplicated in other advertising media, such as newspapers and magazines, and enlargements were displayed in the store, so that images and contexts simulated each other in all modes.

Retail Chain Stores

It is now common to see several branches of famous department stores, but many of the largest chains of stores did not begin as a main store with branches. A chain store system can be designed from the very start, because it is "nothing more . . . than a method of distribution involving the use of more than one retail outlet."[22] The oldest, and still among the largest, is the Great Atlantic and Pacific Tea Company, which began in the 1860s. Soon to follow were the Jones and Brothers Tea Company (later to become the Grand Union), the F. W. Woolworth Company, and the J. C. Penney Company. The term *department store* still tends to connote specialization in clothes and durables, whereas the term *chain store* can refer to stores that provide those commodities as well as to stores providing virtually all other commodities, including pharmaceuticals, lodgings (motel chains), and food. Food chains include not only supermarkets but fast-food restaurants, such as McDonald's, Hardee's, Taco Bell, Pizza Hut, and so on. Department stores and chain stores both depend on a highly efficient and extensive infrastructure of mass production and transportation, but until recently, there was a difference in their geographical concentrations: department stores were located in urban centers, and chain stores, especially fast-food restaurants and chain motels, were geared to automobile life and located along major traffic routes or shopping strips.

As with department stores, chain stores, as places, exhibit the tensions of commodities. They attempt to provide standard commodities that are at the same time distinct from those of their competitors. Commodities must be both generic and specific. When a commodity first becomes available in a mass-produced form, it has the advantage of predictability that comes from standardization. We know exactly what each instance of this commodity is like. One of the earliest food chains, White Castle, capitalized through their advertisements on the fact that their product was generic and predictable: "When you sit in a White Castle remember that you are one of several thousand; you are sitting on the same kind of stool; you are being served on the same kind of counter; the coffee you drink is made in accordance with a certain formula . . . even the men who serve you dress alike. They are motivated by the same principle of courtesy."[23] Needless to say, such a claim holds no attraction now, with hundreds of other fast-food chains.

For chains of gas stations, motels, and fast-food restaurants, the appearance of the place from the highway becomes its sign or emblem, and so the tensions between generic and specific are played out even in architecture. Until the late 1950s, most businesses along shopping strips were not controlled by chains but, rather, were run by local entrepreneurs. The look of the strip was far more varied and irregular.[24] Since then, the chains have risen to prominence, and the strip is neater and far more standard. Buildings are roughly the same size, and particular chains have particular facades. Part of this standardization is due simply to the means of transportation. The strip depends on automobiles, and these have transformed the structure of roads. According to Edward Relph,

> since 1945 highway planning and engineering have greatly expanded a landscape that had begun to be made in the 1930's, a hard-wearing and relatively featureless landscape oriented exclusively to machines and machine speeds. . . . One of the more insidious results of all of this has been the demise of the street. We can tell from the mixed-use streets that still remain that they can serve a variety of formal and informal community needs, including shopping, parades, strolling, traffic access, demonstrations and chance meetings with friends.[25]

It is not that less use is made of these high-speed roads designed explicitly for automobiles but, rather, that they are used for a narrower range of activities.

These roads have standard grades, standard widths, standard lights and signs, standard curbs, standard textures, and even standard colors or shades, a uniformity that makes the roads blend into the background.[26] Driving on them at high speeds blurs the details along the sides, so that the look of a particular chain of gas stations, fast-food restaurants, or motels must be standardized enough to be immediately recognizable. They must also be regularly spaced so that they are accessible to fast right turns. Our familiarity with these places also increases the sense of standardization: images of these structures are constantly encountered in print advertisements and on television. As Relph says, the strips changed "from competitive chaos in the 1950's . . . to the orderly and modulated 'television roads' of the 1980's, in which a relatively small number of corporate outlets compete tastefully and repetitively." Each chain building is an advertisement for the commodity it sells. Some claim that the strip itself is a giant advertisement. The skylines of some cities—like Las Vegas, Nevada—consist not of buildings or trees but of signs and advertisements.[27]

As these places become generic, they must strive to appear specific. Arriving in one of them after a long and tiring ride reassures us by their familiarity, but that may not help us decide which chain to select. Chains have now adopted several facades; McDonald's offers country French, English Tudor, Mediterranean, and dozens of others to pique interest and also to comply with the regulations of municipalities that want facades to blend with local architecture.[28] Chains further their specificity by making it clear that the context their commodity provides is found only at places where they are sold. When MacDonald's says "we do it all for you" or Burger King claims that "we do it the way you do it," their ads make abundantly clear that what they do takes place only within their physical structures. Even the management textbook of the staid and practical National Institute for the Foodservice Industry claims that a restaurant sells not only food but also context: "A foodservice operation's products are more than just its food. They also include service, decor, table appointments, credit policies and the operation's prestige as well as any other characteristics designed to produce consumer satisfaction."[29]

The same tensions on the landscape are found in the downtown sections of major cities. A cityscape can be "read" as a series of signs that display the purpose of its structures. Signs can mean literally those affixed to buildings, but they can also mean the oblique statements made by a building's facade: the architecture, color, and ornamentation of banks, civic buildings, factories, offices, and department stores state their function.[30] We do not usually expect to find pink banks decked with neon lights.

Many changes have taken place in urban centers, including the departure of manufacturing and middle-class residents. As a result of these factors and the increase in concentration of corporate capital, the texture of the downtown now appears more homogeneous and on a much coarser scale. Many small, independent businesses with varied facades have been replaced by huge corporate structures, each one embracing virtually an entire block. As Relph claims:

> By their great size office towers indicate wealth and economic power. They are the cathedrals of the modern city. They are also the corporate flagships, often displaying the company name or logo on high, occupying prominent city centre locations, striving for architectural prestige within the narrow range of high-rise modernism. . . . When conditions are entirely propitious a corporate tower can become an urban landmark, a subtle and prestigious form of advertising. . . . [These messages are] most easily absorbed at a distance, especially from an expressway. . . . At street level the chief impression is one of monolithic austerity. . . . This visible change has been accompanied by a shift of activity from outside on the street to inside in offices and enclosed shopping malls.[31]

We do not come to corporate headquarters to purchase commodities. If we do not work in the headquarters, we have no business there. We see it only from the outside as we pass by. These structures, then, do not have to represent the context of their commodities. Rather, they symbolize the strength and solidity behind these commodities. This is why the Sears Tower does not look like a Sears store.

Shopping Malls

Originally, department stores were located in urban centers, and chains were found mostly along shopping strips and in the suburbs. Now, however, department stores, chain stores, and almost any other kind of retailing have come together in shopping malls. These are mostly in the suburbs, although they have been introduced into city centers as key components in downtown revitalizations. Shopping malls are only one kind of mall. I envision megamalls, that might combine, in a gigantic, insulated, and totally planned environment, the conventional shopping mall, sporting arenas, hotels and motels, and adult and child entertainment centers. These megamalls will be the future "palaces of consumption." Places such as Calgary, Canada, and Edina, Minnesota, have taken steps toward creating megamalls, and the revitalization projects of several major North American cities, such as Boston, Baltimore, and Minneapolis, are like megamalls. I consider first the ubiquitous shopping mall and then the possibilities of the megamall.

The geography of shopping malls differs in several respects from that of ordinary shops on a public street. Along the latter, private businesses bid for store space in an unplanned and uncoordinated system. In contrast, shopping malls are for the most part privately owned and planned from the bottom up. Developers or managers rent spaces to stores under a far more coordinated plan than exists to control shops along public streets. This plan regulates type of store and, especially, size and spacing. Driveways and parking lots often separate mall stores from public thoroughfares. All enclosed malls, and even some open ones, back onto these streets. One of the earliest open malls was built in 1931 by Hugh Prather of Dallas, who had the mall's shops face inward, away from the road.[32] In this design, parking lots and sidewalks are removed from the public domain.

The differences between streets with their stores and malls with their stores are even more marked when the latter are completely enclosed. One of the meanings of the term *mall* is a partially enclosed and sometimes sheltered walk or promenade. (The earliest seventeenth-century references to the word refers not to walks but to part of a polo field.) The enclosed and even multistoried mall is now the norm. It accentuates the principles that operate in department stores. Instead of one large department store with several departments, the enclosed mall houses many stores, some of which might be department stores with several departments. Yet there is an organization among them, for, within certain limits, the mall is centrally planned and coordinated.

The enclosed mall "focuses attention inward" by removing us from the outside world and placing us in a completely artificial environment.[33] Malls have only a few entrances, usually doors with heavily tinted glass so that the view outside is obscured and darkened. Mall stores normally have no outside windows. Artificial illumination is the mall's primary, if not only, source of light. Temperature and humidity in the mall are thermostatically controlled, and once inside, we can lose connection with the cardinal directions. Consumption becomes the mall's compass, and the primary points of orientation are the principal promenades and principal stores and their displays. The mall further enhances the sense of being a world apart through the placement along the promenades of fountains, shrubbery, palm trees, simulated lava, and waterfalls with rocks. Surrounding these oases are little paths with benches, sculptures, and street lamps to offer the shopper repose in this paradise of consumption (see figures 33 and 34). The purpose of the environment is not relaxation, however, but titillation. The mall is there to stimulate the desire for commodities.

As with the department store, the mall and its floor space act as a series of advertisements for its commodities. The enormous array and juxtaposition of commodities imply that one commodity begets other commodities. Our appetites are further whetted by commodities displayed in contexts that make the power of a commodity more explicit: male and female mannequins in trendy clothing and jewelry and sporting the latest hairdos show that the tennis rackets they hold are the latest in fashion; nearby, a leather chair exhibits traditional values and taste, because it is surrounded by bookshelves filled with facsimiles of belles-lettres.

Mall design strives to attain the optimal balance of exposure and variety. The internal partitions of malls are flexible and can be arranged to accommodate new tastes and create new contexts.[34] Considerable research has been undertaken to discover the most effective combination of store sizes and spacing. For example, malls with few stores, each with sixty-to-eighty-foot window displays, are less stimulating than malls with more stores and a twenty-to-forty-foot window frontage.[35] The greater the number of contexts, the better.

As we enter a mall, we might have particular commodities and destinations in mind, but malls assume that, even so, we will browse. Many of us might go to a mall primarily to window shop. The flow of traffic is thus determined by the power of displays to attract our attention. In many malls, the middle point and end points of the promenades are anchored by large department stores, which have the greatest variety of commodities. Between these are smaller, specialized stores. Apart from the anchor stores, which may be starting points for consumers who have a particular product in mind, the mall presents no necessary order or hierarchy to places. Each store, each display, each commodity, cries out equally for attention, and each draws us by promising to resolve the tensions of modernity. They offer to connect our private worlds with the public world of display and to provide us some combination of generic/specific, apart/together, we/they, and self/others. We move along the promenade, stopping here, going there, drawn by whim or random encounter. To the browser, the mall is a world without hierarchy, direction, and necessity.

The reflection of one context by another (note how often the walls of mall stores are lined with mirrors) is most vividly illustrated by a 1989 traveling furniture exhibit sponsored by the Furniture Information Council and hosted by many malls across the country. One of its exhibits, called "Furniture Magic," consists of a bench and a large cabinet containing a closed-circuit television screen about five feet away and facing the bench. The screen forms the upper part of the cabinet; from the lower half, beneath the screen, a camera lens is focused on the bench. On another side of the cabinet, printed instructions explain that this device can visually simulate how we would look in over forty different rooms and furniture settings, including French provincial and Danish modern living rooms and colonial American dining rooms. To produce this illusion, we first must press a button assigned to one of forty furniture contexts, then sit on the bench facing the screen and camera, and then press a button on one side of the bench to activate the camera lens, which is directed at us on the bench.

The television camera portrays us on the screen, but the bench does not appear. Rather, the camera superimposes us onto the chairs or couches that form the setting on the screen. We can even move from chair to chair in the setting portrayed on the screen by moving along the bench. Through these superimpositions of images, we become the center of the ad. A touch of another button produces, within seconds, a print of any of these superimposed pictures, so that in the privacy of our own home, we can contemplate how we look in over forty stage settings, any of which could become ours.

The stores and promenades are the mall's front stage. As with department stores and retail chains, the backstage of planning and work, of the geography and history of products, do not intrude. Commodities seem to arise effortlessly out of nowhere, in character with mass-produced commodities themselves. Advertising, whether in print or through the places of consumption, disguises the threads of meaning, nature, and social relations that are woven into the commodity. But malls are supported by extensive backstage planning and effort. They contain storage rooms, warehouses, plumbing and electrical conduits, heating and cooling systems, managerial offices (up to 20 percent of total mall space),[36] janitorial space, security offices, and employee lounges. Parking is another backstage activity. In malls, the number of slots are carefully calculated according to the estimated volume of business. And malls, of course, rely on consumers having automobiles and disposable incomes and on an unspecified worldwide system of production and distribution. These backstage functions depend on a public, objective meaning of space. They demand that we consider the world from a view from nowhere.

The front stages of malls and other places of consumption, like commodities themselves, combine elements from public, objective space with elements from private, personal place. Malls want us to feel at home, surrounded by friendly strangers. They are public places, privately owned. Indeed, many of the older functions of city centers, public streets and paths, town squares, and village fairs are reworked within the mall's environment. Malls host arts and crafts fairs, orchestras, bands, and dance ensembles; they provide teenagers places to "hang out" (one study shows that teenagers spend more time in malls than any other place except home, work, and school).[37] And the promenades of malls, marked off in fractions of a mile, offer walks for the elderly.

Being themselves displays and advertisement, malls provide an ever-changing scene or spectacle. These characteristics are reinforced by the interplay of media. Malls appear in commercials and as backgrounds for plots in television and movies. Virtually all of the commodities advertised on television are available in malls. Malls sell television sets and often house movie theaters.[38]

Malls, then, are themselves commodities molded by the tensions of the commodities they sell. They are not isolated places but are part of a vast network. Since the 1970s, malls have attracted adjacent activities, such as housing complexes and professional services.[39] They compete successfully with urban downtowns for businesses and shoppers and form nuclei for new types of settlement agglomerations. Malls, in fact, are literally on the map and have become place names.[40]

But malls are not just a suburban phenomenon. Megamalls, composed of interconnected hotel complexes linked to enclosed multistoried shopping complexes, which often form several of the lower floors of the hotels, have become key components in the revitalization of downtowns. The fact that these are consumer-oriented establishments that conform to a commercial plan and that are interconnected by pedestrian walkways makes them mall-like. These complexes are in close proximity to other places of consumption, such as theaters, symphony halls, museums, and metropolitan arenas hosting sporting events. Surrounding these huge places of consumption are workplaces: modern office buildings and corporate headquarters, which form the city's skyline.

Prime examples of this kind of urban revitalization are the Boston waterfront and the inner harbor of Baltimore. These were undertaken as partnerships between the public and private sectors and were based, in Peter Hall's words, on

> a frank realization that the days of the urban manufacturing economy were over, and that success consisted in finding and creating a new service-sector role of the central city. Bored suburbanites would come in droves to a restored city that offered them a quality of life they could never find in the shopping mall. Yuppies . . . would [enter] . . . the blighted Victorian residential areas close to downtown, and inject their dollars into restored boutiques, bars and restaurants. Finally, the restored city would actually become a major attraction to tourists, providing a new economic base to the city.[41]

Government and business coalitions to revitalize downtowns had formed before, but what was new about the ventures in Boston and Baltimore was their huge geographical and financial scale. In the case of Baltimore, ten times more money came from the public than from the private sector. But the point is, the combination of hotels, recreation and culture centers, shops, private housing, intermixed in and around restored and remodeled historic structures recast the entire city as a stage or spectacle.[42] "The process of creating successful places is only incidentally about property development. It is much more like running a theater, with continually changing attractions to draw people in and keep them entertained."[43] The city as stage or spectacle combines the magical features of a mall with those of Disneyland:

> Like theatre it resembles real life, but it is not urban life as it ever actually was: the model is the Main Street America exhibit which greets entering visitors at the California Disneyland, sanitized for your protection . . . wholesome, undangerous . . . Around it, the charmingly-restored streets—all yuppified with a massive injection of HUD funds: they manage to look like a Disney movie lot of an imagined urban America, but they happen incongruously to be real.[44]

•

These urban centers and even larger areas that have become landscapes of consumption exhibit the contradictory qualities of commodities. They can be stimulating, entertaining, and magical, but they can also be superficial, weightless, and inauthentic. These landscapes appear ever more like theater or spectacle as their scale and interconnections increase from advertising individual products, to stores and department stores, to malls, and then to cities themselves. Increasing the scale also increases the size and types of backstages. For individual commodities, backstages are the threads of nature, meaning, and social relations that were involved in their production and distribution. Places of consumption such as hotels, motels, department stores, chain stores and malls assume these particular backstages of the commodities they sell. But they also possess other backstages. One of these is the hidden infrastructure of the workers and managers who undergird these retail and service centers. Yet another backstage appears when these landscapes of consumption become so large and interconnected that they occupy the sites of preexisting centers of production and consumption, as in the cases of urban revitalization. The backstages of these larger places of consumption also include the businesses and inhabitants displaced by the enterprise, who were disconnected from their social networks and their realms of meaning.

Home as a Commodity

The world of consumption is fueled by insatiable appetites for commodities. These desires arise less from the concrete uses to which commodities are put than from our desire to find and express our personal and group identities in a world of strangers.

I noted that there is a history to the development of a private, subjective sense of self occupying an inward mental life and that this is related to an outward geographical segmentation of space.[45] The sense of self as a complete and whole entity may have had its fullest expression in the middle of the nineteenth century, as the public world became more alienating and people sought havens by developing complex private domains.[46] The twentieth century finds an ever-increasing preoccupation with self. Indeed, the children of the latter part of this century have been called the "me" generation, but self, and our relationships to groups, have become ever more fragmented. Commodities have contributed to both the preoccupation with and the fragmentation of self. A sense of self is at the core of the view from somewhere. We cannot escape ourselves and our particular views.

The view from somewhere travels with the self, but most selves have homes. Home does not have to be any particular place or physical structure; home is a place where we are at ease and can let our guard down. As the public realm has become more difficult to share, we literally do find ourselves more at home in the private realm. The norm is to have our physical homes provide this context; although we can feel alienated even at home, this feeling is seen as out of place there.[47] The idealization is of home as a haven from a heartless world, where the self can develop. Recent in-depth interviews have shown that people see their houses as more than investments: they are places that help establish roots; they are refuges that permit us to develop and be ourselves; they are havens where we are in control; they offer the freedom to be responsible for ourselves and yet the freedom from the responsibility to account to others.[48] The association of self with home is reinforced by the fact that, in a consumer society, the home is the primary repository of commodities used to define ourselves and separate our private world from the public world. Western culture has made the connection between home and self-identity more intense for females than for males. At the turn of the century, women were equated with home to such a degree that they were thought to be "homefull," in the word of a turn-of-the-century writer. "Most folks think vaguely of home as meaning marriage, husband, wife, children; but for me its foremost and most beautiful human necessity is a woman; and, indeed this is her finest nobleness, to be homefull for others."[49]

The physical structure of the home provides a setting for the contexts we create with our commodities. We can arrange our furniture, paintings, knickknacks, and carpets to create any context we wish. Though these can be extensions of ourselves, they are not completely subjective if they contain mass-produced and advertised commodities, which embody public meanings. Visitors to our homes see our settings from the perspective of the public language of consumption, and even we might view our contexts and selves this way.

The home and its contents can express our relationship to a group (we/they) and our own personality (self/others). In a study of upper-class Vancouver housewives, Gerry Pratt shows that women who identified with a clearly defined social group used their homes to reinforce this identity and did not see their homes as vehicles for self-expression. Those who had little group identity used their homes and furnishings as a means of personal expression.[50] Still, both groups made use of interior designers, a profession that espouses the view that homes and home decorating should express the occupant's personality. According to interior designer Billy Baldwin, "All rooms should begin as an outgrowth of your own personality, then become more personal, and more welcoming with every year. . . . Nothing is interesting unless it is personal."[51] Purchasing the services of a stranger—an interior designer—to express the real you adds an odd twist to the meaning of commodities and their relation to self-expression.

By joining the private (subjective) with the public (objective), commodities transform the home into a stage or setting to express ourselves (often only the female self) and also into something of an advertisement, resembling a miniature department store. Magazines such as *Better Homes and Gardens, Good Housekeeping,* and the *New York Times Home Furnishing Section* display the interiors of real homes as models of what a home could be like; manufacturers recast these interiors into commodities that can be sold back to the consumer. Although homes are not intended to sell products, they can be interpreted as stage settings, resembling small department stores. Limited resources restrict our purchases and keep our homes far more modest than department stores. But suppose we were incredibly wealthy and had no restrictions on what we could buy or on the size of our homes, which could be the size of castles. If we could buy anything, virtually any activity could be understood as shopping. What would happen? Would we exercise restraint or fill home after home to the brim? What is it about us that could be expressed through such abundance and variety? This might appear to be pure fantasy, except that there are a few who have been near, if not exactly, in this position. In an article called "Lifestyles of the Rich and Tyrannical," Judith Goldstein considers the spending habits of two wives of dictators, Imelda Marcos and Michele Duvalier. She compares them to both ordinary people and other extremely wealthy people whose wealth and power represented a longer tradition and a national history, such as the Shah of Iran.[52]

The palaces and treasure rooms of the Shah were full of the old and the new, the valuable and the ordinary, gaudy baubles and works of art. The rooms were more like a museum than a department store, for each item had a place in the nation's history: one item might be a tribute from a suzerain, another a souvenir of a state visit. The storerooms and palaces of the Marcoses and the Duvaliers also were a melange of items, but their incongruity was more apparent—because, as the possessions of nouveau dictators, they contained little or no tradition or history. In the Malacanang Palace, the Marcos residence in Manila,

> the redwood-panelled living quarters were a blend of opulence, romanticism and vulgarity. . . . Shelves held gilt-framed pictures of the Marcoses with foreign leaders. On the walls hung romanticized portraits of Mr. and Mrs. Marcos. Antique glass cases were filled with jewels and ornaments, mother-of-pearl boxes, ivory carvings, silver bowls, and a plaster Cupid adorned with pearls and diamonds. Built-in shelves were stuffed with unopened boxes of watches, cassette tape players, clocks and pens. . . . Walk-in safes

> [were] filled with crates of jewels and other treasures. . . . Another alcove, mirrored and indirectly lit, was filled with cosmetics and bottles of perfume the size of gasoline canisters. . . . One floor below, Mrs. Marcos kept her outfits, rack after rack of dresses, nightgowns, furs, pants suits, coats and gowns, like the store rooms of a New York department store.[53]

The lifestyles of the rich and tyrannical provide us with several lessons. They show that there seem to be no limits to satisfaction through commodities[54] or that, if there are, it is vulgar and extravagantly ostentatious to push them. Even though they are not Marcos or Duvalier, modern consumers are encouraged to have unbounded appetites. Compared to those in grinding poverty, American middle-class consumers do purchase enough things to have their lives appear extravagant and ostentatious. The consumption patterns of the rich and tyrannical also reveal that the more affluent we become and the greater the volume of our purchases, the more likely the fantasy of consumption will outrun our ability to digest and enjoy what we purchase. Modern advertising draws attention to the power of the commodity. We can be taken in by these ads and become intoxicated by the power of consumption, in general. But commodities require time to enjoy. We must unpack them, examine whether they fit in with the other items we own, or plan how they will form the nucleus of a new context. It also takes time to appreciate the new context they create. If we cannot take this time, the commodities are, in a sense, unconsumed (or literally unpacked), and we are living with only the promise of the advertisements.

Not having enough time to enjoy their commodities might be thought of as a problem of the extravagantly wealthy who are overwhelmed with possessions. But the problem is most of ours: Credit Cards Anonymous reveals that many people are compulsive buyers. They not only are in debt, but they often store away their purchases unopened. In this sense, they have lost control: commodities become burdens when they do not live up to our expectations and yet are not disposed of. These burdens may ruin contexts and make our homes alienating places.

Commodities bring the public realm into our homes, making them less private and also putting their occupants less at ease. Our homes are then displayed and judged. Even if we do not intend our furnishings to make a statement about us, others may still use these items to evaluate us. Ironically, some domestic products are advertised as a means of making the home less a place of repose than a place of factorylike efficiency. The kitchen, for example, is often portrayed as a workplace—a place of production rather than consumption. The homeowner then becomes a worker as well as a consumer, and home becomes a workplace rather than a haven of repose.

Commodities change contexts. Commodities such as radio and television provide built-in systems of continuous change. These media bring into our homes continuous context changes: dramatic, comic, realistic, both the contexts of shows and the contexts of advertisements. Others share these contexts at virtually the same instant, thus further interweaving the public and the private. Moreover, radio, television, video cassette recorders, and stereo systems vary the contexts of consumption at such a rate that the rapidity of change in mood or character of the setting becomes a source of entertainment. Department stores and malls sell these and other pieces of electronic equipment as packages and call them "home entertainment centers." These centers make the home a stage, and the use of home slides, home movies, and home videos allows us and our contexts to be the subject of our own home entertainment. We and our homes can be viewed by us in our homes.

Not only are the interiors of our homes filled with commodities, but our houses and lots and neighborhoods are also commodities and, increasingly, parts of a package. Purchasing an address, or having our homes and surroundings be marks of respectability, has long been important to the way we present ourselves. Although all cultures use artifacts as signs of meaning and status, our consumer society employs different rules to assign these meanings. As early as the 1850s, house plans conformed to particular walks of life. O. S. Fowler designed an octagonal house plan that could be modified to correspond to different occupations or characters.[55] Later in the century, Morrel produced standardized housing in Spanish, Dutch, colonial, oriental, New England, and craftsman styles simply by taking a general plan and altering only the roof, doors, and windows.[56] And in the early twentieth century, Sears and Roebuck published a catalogue called *Homes of Today,* which displayed their ready-made, mass-produced houses.[57]

With further developments in production and advertising, the mass production of homes as contexts has accelerated to the point where new housing developments spring up virtually every minute. With equal rapidity, old developments lose their meanings. Thus the making of houses into images has accelerated and the geographical scale of these endeavors has grown. With the purchase of a new residence, we can acquire a completely designed neighborhood, which includes coordinated house styles, landscaping, recreational facilities, schools, shopping centers, and perhaps even security patrols. Such ready-made communities can be organized around themes—architectural and stylistic, or recreational. Edward Soja gives an example:

> [T]he new town of Mission Viejo . . . is partially blocked out to recreate the places and people of Cervantes's Spain and other quixotic intimations of the

> Mediterranean. Simultaneously, its ordered environment specifically appeals to Olympian dreams. Stacked with the most modern facilities and trainers, Mission Viejo has attracted an elite of sport-minded parents and accommodating children. The prowess of determined local athletes was sufficient for Mission Viejo to have finished ahead of 133 of the 140 countries competing in the 1984 Olympic Games in the number of medals received. [The community is] advertised as "The California Promise" by its developer, currently the Philip Morris Company.[58]

The idea of a settlement as theme park is found in retirement developments. Some are organized around fishing and boating, others around golf. Nearly all of the Delray Beach area of Florida is composed of retirement condominiums surrounding dozens of eighteen-hole golf courses. These communities usually do not permit families with children. (And because of their specialized nature, it would be difficult to rear children in these communities even if they were allowed.) The shape of these golf courses is like a figure eight, with nine holes in one loop and nine in the other. At the intersection of these loops is the clubhouse and swimming pool. Encircling the two loops is a road, off of which are the parking lots. Beyond the parking lots are the clusters of one- and two-story condominiums that house the residents. Surrounding the condominiums are major roads that intersect at right angles and connect one golfing community to another. Along these major roads are the shopping malls and department stores that provide retail services for the communities; because of the size of golf courses, these shops are accessible only by automobile. One single theme draws the elderly residents of these communities together—golf.

Commodities help us express ourselves. The more commodities we consume, the more selves we express. The multiplication and fragmentation of self occurs not only through a proliferation of commodities but also through a proliferation of places of consumption, and even of homes. A home can be a house we own, or an apartment we rent, or a condominium we own whose grounds are shared.[59] Added to these options is time sharing, which allows us to own or rent an apartment, house, or campground for a specified period of the year. We can mix and match these to extend our number of homes. In addition to fixed sites, there are campers, caravans, and mobile homes. Each of these options can place us in radically different contexts, which permit us to express different parts of ourselves or escape from our normal home and context.

As the number and types of living accommodations proliferate, and as they promise to offer contexts for self-expression and fulfillment, we approach the position of having our homes become places that can be used to escape from work and also from our other homes. If we have the option of taking a vacation from our home in another home or in a time-shared condominium, then even our own residences become like tourist places. Indeed, consumption puts us in the frame of mind that there is virtually no activity, including dwelling, that is not like shopping or, in this case, like shopping for tourist attractions. Consider a breathtaking view. A common reaction is, if only we could have a cabin or summer place overlooking it. And a common result is that such a place is either bought privately for the pleasure of a single family (see figure 23), or becomes a resort that rents space so that we can consume the landscape. But before we consider how travel and tourism compare to dwelling and how they create landscapes of consumption, we should remind ourselves again of the ubiquitous but hidden backstage—the world of work.

The Workplace

The world of consumption rests on the world of work, and much of that world employs a sense of space and place that is practical and distant. Business and commerce see the world in terms of public (objective) space. To plan the location of industries, the circulation of raw materials and finished goods, the construction of homes, department stores, and shopping malls, and the development of tourist centers assumes a perspective from nowhere. This sense of space has already been addressed and most works in the geography of production and consumption fill in its details. Here I will discuss another facet of work.

Work is often seen in opposition to being home or being on vacation. This opposition is complex. First, work provides the means for us to consume, to purchase a home, and affords us the means for entertainment and travel. Most consumers are also workers. Second, work builds and sustains commodities and places of consumption; work is the backstage of consumption. Third, work, being often disagreeable and alienating, needs to be escaped via consumption, home, and travel. Since we do not express ourselves in work or find work fulfilling, we need to search for self-expression through consumption. The first point is more or less self-evident, and so I turn to the latter two, beginning with the third.

Alienation is a term often associated with modern work. Work can be alienating in at least two closely related ways, stemming from the fact that workers do not control the means of production (which can happen in economic systems other than capitalism).[60] First, work is not defined in terms of a meaningful process over which we as workers have control. Tasks are often so repetitive and discrete that they are boring and seem pointless. We do not know how our contributions fit within the larger production process. Second, our work is under the supervision and control of others. This lack of control makes us guests in someone else's place. These two problems are related: the minute specialization and triviality of tasks can be maintained because the work environment is not in the hands of the workers. In some cases, alienation of space may have even preceded alienation of tasks. For example, in the handloom weaving industry of eighteenth- and nineteenth-century England, entrepreneurs collected these looms and housed them in their sheds and hired weavers to work for them and under their supervision. Weavers who possessed their own looms in their own cottages resisted such work because of their loss of control over working conditions, which came with a loss of control over the workplace.[61]

This idea of alienating work can be stated more generally by connecting it to the problem of forces and free agency. The assumption seems to be that nonalienating or fulfilling work involves creating our own projects over which we have control. This means being free not only of particular forces of social relations but also of the arbitrary forces of nature and meaning. Setting our own tasks and fulfilling them means that we are not agents of other forces but agents in our own right. In addition, it assumes that we possess such interests and have the skills to carry them out.

By extending this line of reasoning, real fulfillment in work can be thought of as the pursuit of a life's project, and this could also mean that such a project is the way life attains meaning. This leads to an image of work quite different from the minute technical division of labor that exists in all highly industrialized countries, even those where workers are technically in control of the work process. Work as a long-range or even lifetime project is also in opposition to the sense that life is a series of lifestyles that can be changed as easily as we change our wardrobes or furniture. And if work is a project that demands commitment because it is intrinsically interesting and self-fulfilling, we would not need to escape work, because we would not need to escape from ourselves.

But the contemporary world is not like that. Few of us are fortunate enough to have sustained interests in projects of any kind, and even if we have, they are difficult to pursue because they rarely mesh with the technical division of work. The result is that work for most of us, although it is perhaps not entirely drudgery, does require escape and the need to feel in control. This need is one of the psychological voids that consumption fills.

Ironically, places of consumption can themselves create the need to escape the drudgery of work, if one happens to be an employee of such a place—perhaps as a salesperson or a janitor in a department store, mall, or resort. These workers are either part of the front stage, or they work backstage to arrange and maintain the front stage. These workers are part of the process that Walt Disney called "imagineering."

The connection between work and escape from work becomes even more blurred when we work at home, as homemakers and as employees who can use modern communication systems, flexible hours, and piece-work to perform the same tasks at home as at the office or factory. Working at home may allow us greater control over some aspects of work, especially its timing, but it does not necessarily give us control over the division of labor or the use to which our labor is put. Working at home may even lead to a further loss of control in the work process by bringing the surveillance and control of supervisors through electronic monitoring systems into our private domains.

Escape from work may also be difficult for employees who go on working vacations or retreats organized by their companies at resorts rented by their companies. These resorts provide distractions for workers and their families and conference rooms and lecture halls to facilitate work. The boundaries between front stage and backstage become even more vague if people who work at a resort go on a retreat in another resort.

Tourism as a Commodity

What is tourism, and how does it differ from exploration and travel? In an article on travel, Paul Fussell presents the generally accepted differences among the three. "Each is roughly assignable to its own age in modern history: exploration belongs to the Renaissance, travel to the bourgeois age, tourism to our proletarian movement. . . . All three make journeys, but the explorer seeks the undiscovered, the traveler that which has been discovered by the mind working in history, the tourist that which has been discovered by entrepreneurship and prepared for him by the arts of publicity."[62] This description of the three is insightful, if we disregard the implied condemnatory note that tourism is somehow the least authentic. Yet to many critics, inauthenticity pervades tourism. Tourism does require the preparation of places, at least so they are accessible. And it does require publicizing the place's qualities. In this sense it is more staged. And unlike tourism, travel, as its etymology suggests is (or perhaps more accurately was, because it is now difficult to be a real traveler) more work: the word derives from the French *travail* meaning "labor, work, toil," which is from the Latin *tripalium,* an instrument of torture.

Daniel Boorstin also makes the point that tourism is ready-made and easy, and that "until almost the present century, travel abroad was uncomfortable, difficult and expensive."[63] Not only was travel more difficult than tourism, it did not focus on the sense of sight. Until the sixteenth century, travel emphasized the auditory: one traveled to discourse with learned people. It was only later that "travel literature began to argue for the superiority of the eye over the ear."[64] Even more than travel, tourism relies on the visual—the scene, the view. Photos and postcards are the tourist's means of capturing the trip and experiencing it again and again.

Creating tourist places embeds the tensions of commodities into the landscape. Tourist attractions can become like other places by providing familiar accommodations and foods. They can become distinct by supporting or reinventing local customs. They can make us part of a group by catering to a certain clientele (see figure 20). They can help us find ourselves by offering us self-absorption or by assisting our escape to an exotic place; through the process of immersing ourselves in a place, we come to discover our true identities (see figures 18 and 19).[65]

But what has been displaced by embedding these tensions in the landscape? Some indication of the process of landscape transformation is available in Walter Christaller's description of a hypothetical tourist sequence:

> Painters search out untouched and unusual places to paint. Step by step the place develops as a so-called artist colony. Soon a cluster of poets follows, kindred to the painters: then cinema people, gourmets, and the *jeunes dorées*. The place becomes fashionable and the entrepreneur takes note. The fisherman's cottage, the shelter-huts become converted into boarding houses and hotels come on the scene. Meanwhile the painters have fled and sought out another periphery—periphery as related to space, and metaphorically, as "forgotten" places and landscapes. Only the painters with a commercial inclination who like to do well in business remain; they capitalize on the good name of this former painter's corner and on the gullibility of tourists. More and more townsmen choose this place, now in vogue and advertised in the newspapers. Subsequently the gourmets, and all those who seek real recreation, stay away. At last the tourist agencies come with their package rate travelling parties; now, the indulged public avoids such places. At the same time, in other places the same cycle occurs again; more and more places come into fashion, change their type, and turn into everybody's tourist haunt.[66]

R. W. Butler has made this description into a formal tourist-cycle model.[67] (The term *cycle* is his, though perhaps *stage* or *sequence* would be more appropriate terms.) The first stage is the discovery of the place by "explorers" (adventuresome tourists). This triggers an attempt by locals to become involved in catering to, and attracting more of, such tourists. This involvement often leads to "some advertising . . . to attract tourists . . . a basic initial market area for visitors . . . and a tour season [which requires adjustments] in the social patterns of at least those local residents involved in tourism."

The next stage is one of more intense development, wherein local investors and facilities are superseded by "larger, more elaborate . . . external organizations. . . . Natural and cultural attractions will be developed and marketed specifically, and these original attractions will be supplemented by man-made imported facilities. Changes in the physical appearance of the area will be noticeable, and it can be expected that not all of them will be welcomed or approved by all of the local population."

The consolidation stage is next, with its increasing emphasis on marketing and the involvement of franchises and chains. After this stage, a period of stagnation can be expected. "The area will have a well-established image but it will no longer be in fashion. There will be a heavy reliance on repeat visitation and on conventions and similar forms of traffic. Surplus bed capacity will be available, and strenuous efforts will be needed to maintain the level of visitation. Natural and genuine cultural attractions will probably have been superseded by imported 'artificial' facilities. The resort image becomes divorced from its geographic environment."

After this stage is decline followed by rejuvenation, perhaps by the introduction of a new attraction, such as gambling, or by tapping a previously unused resource, as when summer holiday villages become ski resorts in winter. This sequence can be found over and over again in landscapes as large as the Florida coast, Hawaii, and the Costa Brava of Spain. The sequence can also be compressed in time, as in the "instant" resort development of Cancun, Mexico, which was selected by computer from a number of competing sites.

The effects of tourism are both positive and negative. A positive side is that tourism provides access to places that otherwise would be enjoyed only by the extremely rich or by those who had the good fortune to live there. And yet this accessibility—to a wilderness area, a ski slope, or a beach—can degrade these places to the point where nothing of value remains. Even attempts to preserve them can do more damage than good. Perhaps the classic example is the accelerated erosion of beaches through attempts to stabilize them.[68]

Landscapes of consumption, then, tend to consume their own contexts. The cycle of accessibility and degradation applies not only to the natural environment but to the cultural context, as well. In cases where the political system is repressive, tourism furthers the injustice by disadvantaging the indigenous poor, which in the long run diminishes the attractiveness of the place.[69] But many cases are more ambiguous. Consider Erik Cohen's analysis of the Alarde, a public ritual in the Basque town of Fuenterrabia. The ritual "became a major tourist attraction [which was so successful that] authorities declared that it should be performed twice on the same day to accommodate the large number of visitors." The performance then became staged, and because of that "the local participants lost interest."[70] To help to restore interest "the municipal government was considering payment to people for their participation in the Alarde . . . just as the gypsies are paid to dance and the symphony orchestra is paid to make music."[71]

Payment might destroy the meaning of the ritual, or it might change or enrich the ritual by the introduction of new elements and levels of meaning. This enrichment can be seen in Balinese ritual performances, which now have three separate audiences: "a divine, a local, and a tourist."[72] Tourists do not necessarily spoil the meaning of the performance for the others. "The touristic audience is appreciated for the economic assets it can bring . . . but its presence has not diminished the importance of performing competently for the other two audiences, the villagers and the divine realm."[73]

Even if the consequences of tourism can be beneficial, it has a homogenizing effect on places and cultures. By their sheer numbers, tourists put a cosmopolitan stamp on the places they visit. For example, the physical structures that accommodate the tourists tend, despite veneers, toward a worldwide standard hotel-motel architecture. The commerce, currency, and language of the tourists also alter the local culture, making it more cosmopolitan, less alien, and less exotic. Generic qualities overwhelm the specific. Consider how much alike are the shores of Waikiki, Miami, and San Juan. We may wonder about the authenticity of such places.

The sheer act of planning and organizing events and engaging in environmental control on a scale large enough to entertain masses of people means that the landscape becomes more and more like a stage. One consequence is that the messy part, the part that distracts from the fun and fantasy, is pushed out of view—to the backstage. When enough tourists wish to see how a people live their daily lives, these backstage activities are pushed to the front. Both lead not only to a more homogenized middle stage, in Meyrowitz's term, but to a problem of staged "authenticity." To Boorstin, tourist attractions provide "local atmospheres" that are pale replicas of the real culture.[74] "These 'attractions' offer an elaborately contrived indirect experience, an artificial product to be consumed in the very places where the real thing is free as air."[75] As Erik Cohen summarizes:

> Contrived cultural products are increasingly "staged" for tourists and decorated so as to look authentic. . . . Above all, tourists, who are apparently permitted to penetrate beyond the "front" areas of the visited society into its "back" . . . are in fact cheated. Such back regions are frequently inauthentic "false backs," insidiously staged for tourist consumption. Thus, for example, localities may be staged as being remote or "non-touristic" in order to induce tourists to "discover" them . . . and native inhabitants of "exotic" places are taught to "play the native" in order to appear "authentic" to the tourists.[76]

Cohen provides several examples of how natives stage ceremonies in even remote areas only slightly touched by tourism. These areas become attractive to tourists—who are now called trekkers, to remind us that they are like travelers, but perhaps more athletic. These places have remained specific and alien enough not to have become comfortable tourist attractions.[77] If the site is remote and virtually untouched, the staging can be seen more clearly in the way the place is advertised ("communicative staging") than in its actual landscape features, or "substantive staging."[78] A case in point is a trekking ad for a guide company that takes tourists to visit a remote hill tribe in north Thailand. One part of the ad reads "join the hill tribe native people dance in their traditional steps at camping." This is communicative staging, because "while in some . . . villages the natives, especially young women, do indeed tend to meet at night spontaneously for singing sessions, in [the village mentioned in the ad], which was for many years the most frequently visited hill tribe village in the standard trekking areas . . . the dancing was performed expressly for the tourists, and paid for surreptitiously by the guides."[79] Cohen observes similar "false advertising" and "staged authenticity" with regard to the food the tourists are served and even the backgrounds of the guides.

The tourist, though, may not be concerned about authenticity. The feeling that landscapes are inauthentic depends on how we evaluate particular parts of the bundle of the tensions of commodities. These differences lend a relativity to authenticity. Anthropologists and geographers may be the ones most negatively affected by inauthenticity; the trekkers may find it stimulating and entertaining, or they may be unaware of any pretense.[80] Some people find that transformation of place is authentic in the context of the modern world. They think of modernization as the dislodging of elements from their original contexts and placing them with other dislodged elements to form a new context.[81] According to Cohen:

> In principle it is possible for any new-fangled gimmick, which at one point appeared to be nothing but a staged "tourist trap," to become over time, and under appropriate conditions, widely recognized as an "authentic" manifestation of local culture. One can learn about this process of gradual "authentication" from the manner in which the American Disneylands, once seen as the supreme example of contrived popular entertainment, became over time a vital component of contemporary American culture.[82]

Umberto Eco finds a twist in the process of authentication. If authentication can make anything authentic, and if our landscape is a pastiche of things borrowed and imitated, then it can create in us not only a yearning for the authentic but a belief that the genuine article can be reproduced. He believes he has found such attempts in his journey through a part of the American landscape he calls hyperreality. He is "in search of instances where the American imagination demands the real thing and, to attain it, must fabricate the absolute fake."[83] Eco's examples of how the "really real is the absolute fake" include "a polychrome Venus de Milo, with arms; bronze casts of Roman reproductions in marble of Greek statuary; wax technicolor versions of Leonardo's Last Supper; a total reconstruction of a buried Roman villa, as it is thought it would look."[84] To these fabrications can be added the theme parks that attempt, through fantasy, to recreate contexts. But Eco overdraws the case when he suggests that there is no difference between the real and its simulation. The meanings of the terms themselves prove the difference.

Still, we do often gloss over these differences and may even be unaware of them. Such can be the case in examples closer to home. For instance, outdoor museums are supposed to capture life as it was at a particular period and place. Yet the museum's artifacts may never have been together in the way they are exhibited and at the same moment in history. Historically preserved or restored areas attempt to show places as they were, when in fact they were always changing. The huge tracts of land that are set aside as wilderness are attempts to preserve pure nature, yet they require constant human intervention to sustain them.[85]

In all of these cases, the essence of the thing—the really real—requires human intervention and fabrication, and the places themselves become commodities. They become tourist places: museums and national parks may advertise and charge admission, and both may become tourist attractions within larger areas, which then advertise themselves as tourist attractions. Cities mention museums in their advertisements; brochures for outdoor museums describe their authentic atmospheres; and regions of the United States, such as the Southwest, incorporate national parks as part of their context.

The issue of place as commodity can be extended to entire countries that are advertised and visited as though they were theme parks. Simply consider the popular images of merry old England, of romantic France, and of clean and efficient Switzerland. These images are drawn from what exists, but are accentuated by the fact that contexts are what tourists expect to see. Tourists want the places they visit to be accessible and not too alien, which means that the places must become generic and yet retain enough specificity to be attractive. These opposing aims lead to certain consequences.

One consequence of a pastiche being seen as authentic or having an imitation or reproduction stand for the real thing is that these pastiches and reproductions can become the elements of yet other images or reproductions. The cycle can accelerate to such a pitch that the referents of the images become blurred and the images themselves appear as reality. In Eco's words, "the sign aims to be the thing, to abolish the distinction of the reference," and things "must equal reality even [if] reality was fantasy."[86] Jean Baudrillard says the same thing when he calls ours an age of simulation, a play of signs.[87]

These postmodern characterizations of the world as hyperreality, as a reality without origin, are hyperbolic, for they take the consumer's world as the only world and deny that we know more than ever about the past and possess a greater understanding of the objective positions and relations of things in the world.[88] Yet they do capture the disorientation that comes about with the commoditization of the world. They identify the extreme problems of commoditization's acceleration of the freedom and burden to create meaning. They recognize that when meanings are clearly established and have unambiguous referents, the symbol system cannot be played with at will, but that when meanings have been loosened or when the things the symbols refer to have lost their standing in a hierarchy of meaning, then they can be used more freely, even to the extreme of conveying no meaning at all. The most complete geographical microcosm of these tensions is in theme parks, and the best known of these are Disneyland and Disney World.

Disney World

The tensions between reality and fantasy, between the really real and the truly fake, come together in the gigantic theme parks. These include Busch Gardens in Florida and Virginia, Six Flags over Texas and Six Flags over Georgia, King's Island in Ohio, Opryland in Tennessee, Santa's Village in Illinois, Astroworld in Texas, and Cedar Point in Ohio. Many of these, though, are smaller versions or imitations of the ultimate in theme parks, Disneyland in California, and Disney World in Florida. Disney's theme parks are worthy of attention simply for their sheer size and audacity of planning. They contain the traditional rides and sell enormous quantities of commodities, but their primary commodities are the landscapes or contexts that we experience. Disney theme parks are completely controlled worlds of fantasy, aptly described as "atmospheric parks"[89] and billed as the "happiest places on Earth."

Disney World in Orlando, Florida, is the largest of Disney's parks. The entire complex, called Walt Disney World Vacation Kingdom, covers twenty-eight thousand acres (forty-three square miles), or approximately the size of San Francisco and twice the size of Manhattan Island. By 1990, the Kingdom had attracted more than 183 million tourists (averaging over 21 million annually), and employs over twenty-two thousand people during peak vacation periods. It is a vast but minutely designed theme park with two major attractions: the Magic Kingdom and the EPCOT (Experimental Prototype Community of Tomorrow). Each has its own entrance and admission, but a special ticket can be purchased allowing admission to both and to free and unlimited use of the connecting transportation system.

The Magic Kingdom, which is a larger model of Disneyland near Los Angeles, is divided into four major parts. One is Adventure Land, and includes among its attractions the Swiss Family's tree house, which beckons you to climb through a giant replica of the shipwrecked family's home, and a Jungle Cruise, which calls for us to board an explorer's launch for a "danger filled" cruise down tropical rivers of the world. The second part is Frontier Land, which contains Big Thunder Mountain Railroad, offering a roughshod roller-coaster-type ride through the days of gold rush excitement, and Tom Sawyer's island, which is accessible by a raft journey across the Rivers of America and contains the rustic Fort Sam Clemens. The third is Fantasy Land, offering a Mad Hatter tea party, a sail with Captain Nemo on a submarine voyage twenty thousand leagues under the sea, Snow White's adventures through the dark forest to meet the seven dwarfs and the wicked stepmother, and numerous fairytale rides and shows. The fourth is Tomorrow Land, which contains futuristic rides and exhibits. All four parts of the Magic Kingdom are approached through the *axis mundi* of Main Street USA, where yesteryear's small town America comes to life through a Walt Disney railroad, store, Main Street cinema, penny arcade, and old-fashioned vehicles.

The EPCOT Center surrounds a lake and displays both technological and cultural environments with supposed educational value. The technological exhibits are sponsored by major corporations and include the Universe of Energy (presented by Exxon); Horizons, an exhibit of the lifestyles of the twenty-first century (presented by General Electric); World of Motion (presented by General Motors); Journey into Imagination (presented by Kodak); the Land (presented by Kraft); and the Living Sea (presented by United Technologies Corporation).

The cultural environments are across the lake and consist of seven national showcases, each attempting to capture an image of a country through architecture, food, music, audiovisual techniques, and shops. The nations portrayed are Mexico, with its El Rio del Tiempo and its restaurants and shops, including the Plaza de Los Amigos, Artesianía Mexicana, and Fiesta Fashions; China, with its circle vision show of the Wonders of China and its restaurants and shops, including the Yong Feng Shandian, containing "authentic gifts from China"; Germany, with its Biergarten and shops; Italy, with its restaurants and shops; the American Adventure, with an "inspirational" story of America, American Gardens Theater, a shop offering antiques and hand-crafted goods and American items, and Liberty Inn, offering hamburgers and hot dogs; Japan, with the Bijutso-Kahn Art Gallery, restaurants and shops; Morocco, with the Gallery of Arts and History, Fez House, restaurant, and shops; France, with a motion picture impression of France, restaurants, and shops; the United Kingdom, with a pub, a dining room, and shops; and Canada, with a motion picture of Canadian scenery, a restaurant, and a shop.

These showcases and entertainment centers in the Magic Kingdom and the EPCOT Center are supported by an equally huge, complex—though, in accordance with Disney's plans, invisible—backstage—for it was always Disney's wish not to have the real world intrude. As he put it, "I don't want the public to see the real world they live in while they're in the park. . . . I want them to feel they are in another world."[90] Gatekeepers, vendors, and grounds keepers, the only visible part of the backstage crew, are all in costume. But the vast infrastructure, the telephone and power lines, transportation lanes supplying commodities and removing refuse, the maintenance facilities for the highly complex and technical forms of illusion—the activities that keep this fantasy afloat—are out of sight. Engineering an infrastructure that has enormous efficiency and complete invisibility is one of Disney's major feats. Paradoxically, this feat is an attraction in itself and even a subject of advertisement, but primarily for city planners and environmental engineers. Its prototype technology for underground power-generating facilities, underground sanitation facilities, advanced telecommunication systems, vehicle guidance systems, and show control and support systems has been the subject of articles in technical journals such as *Industry Week, Engineering News Record, Electrical Construction and Maintenance,* and *Concrete Products.*

So important is the sense of fantasy set apart from the real world—a front stage without a backstage—that the Disney manuals for instructing workers use a dramaturgical vocabulary to describe the entire project. Employees do not have jobs but "roles." They do not apply for jobs but "audition" for "parts." The "cast" wears not uniforms but "costumes." The Walt Disney World Vacation Kingdom is not a business, but a twenty-eight-thousand-acre "stage," and the public are not customers but "guests," composing an "audience" who are all on "stage."[91] And, to ensure that audience and actors mix, one manual warns, "one should be careful never to allow a show to become a monument."[92]

Once customers pay admission, they enter a paradise of entertainment where practically all of the attractions of the Magic Kingdom and EPCOT are free. The only constraint is time, which is measured in length of lines and distances from one stage to another. Unless one remains for weeks on end, it is impossible to experience everything. Money does become important for food and refreshment, and for the enticements in the ubiquitous shops selling souvenirs, cosmetics, and books and in the specialized shops that are part of the local environments of the EPCOT Center's cultural exhibits. Culture is presented mostly as a commodity; a journey to a foreign land, like Mexico or Morocco, becomes a shopping expedition.

The power of Disney World is enhanced by the fact that it contains other places of consumption. In addition to the Magic Kingdom and the EPCOT Center, Walt Disney World Vacation Kingdom contains River Country, an "updated version of the 'ol swimmin' hole'," complete with water wings, sandy beaches, inner-tube rapids, and a twisting, 260-foot water slide; Discovery Island, a certified zoological park, with wilderness trails, exotic plants, and wildlife; lakes for swimming, boating, fishing, and water skiing; a seventy-five-hundred-acre wilderness, which has been set aside as a permanent conservation area; an airport; Walt Disney Shopping Village, featuring twenty-three shops, eight restaurants and snack bars, and three marinas; six hotels; a thirty-six-hole championship golf course; four Walt Disney World resort complexes, consisting of Contemporary Resort (1,050 rooms), Polynesian Village Resort (855 rooms), Golf Resort (288 rooms), and Fort Wilderness Campground (727 camp sites and 463 rental trailers); six Walt Disney World Village Resorts (585 family units); and Walt Disney World Conference Center (accommodating up to 300 people).

Disney World banishes the backstage from view. Even when the park hosts the working world through conferences at one of its conference centers, work is supposed to be tempered by the fun and fantasy of the park. This mixture of reality and fantasy applies also to the acquisition of knowledge. Walt Disney believed that fun and entertainment could be educational and that education should be fun. The presentation of knowledge in the EPCOT Center attempts to adhere to these principles. The exhibits serve as instructional stages or giant visual aids, and the center also contains a huge complex—the EPCOT Teaching Center—to distribute educational materials. The Teaching Center has hosted over eighty thousand educators and sells an array of educational materials to supplement the exhibits, including complementary study guides for all of the exhibits; filmstrips on planetary science, earth science, and careers in science; the twenty-page *Disney Software News* issued several times a year and containing hundreds of Disney software programs on a variety of subjects for elementary and secondary schools; an educators' idea exchange network; a pen pal program; and a means of offering school credit to students who visit the EPCOT Center.

Disney World is above all a business. It transforms knowledge, work, and fun into commodities and thereby infuses these things with the tensions of commodities. It draws things together from different times, different parts of the world, and different domains of reality and thereby juxtaposes reality and fantasy, past and future, and distant places and near ones. It makes generic amusement park rides seem specific by presenting them in new contexts. But it also makes the specific more generic: France and China, Mexico and Morocco become shopping malls, and even Disneyland is replicated in the Magic Kingdom of Disney World and in other Disneylands near Paris, and in Tokyo. (Tokyo's Disneyland contains virtually no Japanese food or signs, but instead of a Mainstreet USA, the entrance is called World Bazaar.)[93] Disney World promises to release us from our own contexts and place us at the center of other contexts: we can be astronauts journeying to Mars, physicists living in a satellite community near the moon, children drifting along the Mississippi River on Tom Sawyer's barge, or forty-niners in the gold rush. It provides us with an opportunity to find our roots, to return home to Main Street USA, and to create a "world of our own." It allows us to distinguish our group from others, urbanites from country folk, Americans from Chinese.

Disney World merges private and public fantasies and makes them palpable as contexts. These contexts can be recreated through other commodities and simulated again and again within other contexts. This simulation is part of the purpose of souvenir shops, film shops, and the commercialization of educational aids. The *Walt Disney World Merchandise Catalog,* available along with the souvenirs in the Disney stores in the park, advertises its products as "capturing the magic and splendor of this fairy tale land."[94] *The Magic Kingdom Guide Book,* sponsored by Kodak, alerts guests to the Camera Center near Town Square and to other sites in other "lands" that also contain Kodak film, film processing (with a two-hour development "just like magic"), camera loans, and a Disney Learning Adventure for Adults that helps "turn your snapshots into unforgettable memories . . . through a workshop called "Capturing the Magic: A Kodak Photography Workshop." Children can make their memories palpable with Disney character dolls. "Cozy up to your favorite Disney characters. . . . Every child needs a friend he or she can hug."[95]

The mixtures and penetrations of public and private, of meaning, nature, and social relations, are multiplied indefinitely. Knowledge is a commodity that can be absorbed with little effort. The Universe of Energy, "presented" (note the theatrical term) by Exxon, distributes a comic book, which provides a tour of the exhibit and instructs children about the nature of energy and conservation through Disney characters Mickey Mouse and Goofy. Disney's *Teacher Tool Software* advertises a language arts program for grades three through seven called Goofy's Word Factory, which has Goofy leading grammar lessons, and another called Mickey's Space Adventure, for grades three through six, which has Mickey Mouse and Pluto teaching children about the solar system. These and other instructional tools are offered to teachers and can be adopted by public school systems. The EPCOT Center itself offers its exhibits as a school classroom, with school credits.

•

Disney World commoditizes practically anything and presents everything at the same accessible level, whether it be learning, recreation, entertainment, commodities, or places such as schools, zoos, or wilderness. It makes the completely artificial jungle containing mechanical animals seem natural, while it creates and manages wild and natural places. It interchanges fantasy and reality. It hides the backstage while pridefully advertising its existence and design. The interplay of contexts and their presentation out of context—the presentation of "instant worlds" in Hidetoshi Kato's term—is the same juxtaposition that we observe in all landscapes of consumption.[96] Such free association is reinforced by connections among mass communications media.

Indeed, Disneyland and Disney World allow us to walk through the stage sets of movies and television. "Disneyland itself is a kind of television set, for one flips from medieval castles to submarines and rockets."[97] And these connected stage sets are not unlike megamalls, residential theme parks, and downtown urban renewal efforts. Disneyland has been called the Town Square of Los Angeles. As with all landscapes of consumption, Disney's theme parks are exciting, entertaining, and mystifying. But they are also scorned as shallow, inauthentic spectacles, which distort the real conditions of nature, meaning, and social relations.

China, Bloomingdale's, and the Metropolitan Museum of Art

The shared tensions of all places of consumption make each a mirror of the others. Los Angeles can be compared to Disneyland, and Disneyland to Los Angeles. These reflections can be initiated by places of any scale, and they jump from scale to scale. A small-scale context can trigger powerful emulations in other places, both small and large. Debora Silverman describes a complex chain of reflections involving principally Bloomingdale's department store, the Metropolitan Museum of Art in New York City, *Vogue* and *Harper's Bazaar* magazines, and China. This chain brings the question of authenticity into sharp relief.[98]

In 1980 Bloomingdale's "unleashed the largest merchandising venture ever in the history of the store by transporting the riches of China to New York." Ads proclaimed that the consumer could visit China without a passport and experience its sounds, sights, and smells. The store offered gold and jade jewelry, bowls and plates, boxes, carpets, jackets, and silk robes with gold threads and sequins. Many of these robes draped over department store mannequins were inspired by authentic Chinese imperial robes on display (but not for sale) in a special museumlike exhibit at Bloomingdale's. The Chinese goods were advertised as timeless, aristocratic, and rare, drawing upon prerevolutionary and preindustrial Chinese styles and skills. They were handwoven, hand painted, hand carved, and in general handmade by artisans skilled in the ancient crafts.

But the fact of the matter was that these were simulated products resulting from an alliance between Bloomingdale's managers and the government of mainland China. China manufactured the goods according to Bloomingdale's specifications. "The items commissioned were designed by Bloomingdale's artists in New York, who projected their fantasies of the opulent, mysterious empire and ordered accordingly." The commodities, though, were handmade in that they were "produced at human assembly lines, where massive orders for hand-painted fans, bowls, plates, and clothes were filled by intense division of labor and specialization." Paradoxically, Bloomingdale's proudly publicized this American-Chinese cooperation in design, which it saw as a means of providing the American public with "authentic" Chinese products to fit American tastes and contexts.

Bloomingdale's effort to promote Chinese products was a success and soon filtered down to other department stores and malls throughout the United States. But Bloomingdale's was not the sole promoter of China. As Bloomingdale's exhibit drew to a close, New York City's Metropolitan Museum of Art opened an exhibit called The Manchu Dragon: Costumes of China, the Ch'ing Dynasty, 1644–1912. The exhibit concentrated on the genuine imperial robes on display for a time at Bloomingdale's. The exhibit was organized by special museum consultant Diana Vreeland, who had been editor of *Harper's Bazaar* and *Vogue* magazines. The museum exhibit was to be a display of only the "rare, authentic" imperial costumes. A museum, after all, is supposed to be the most authoritative source for determining what is or is not authentic. Yet "the ethos of the Metropolitan's curator [Vreeland] was closer to a marketing strategy than to the task of historical education," and so the exhibit became much like a department store floor advertisement. The robes were displayed on department store mannequins, their labels contained information about the fabrics but very little about the dates and uses of the costumes—which were not intended for "public display, but for private worship." The mannequins draped with their robes stood in a dimly lit gallery with "high-pitched ancient Chinese music" and a scent designed by Vreeland's friend Yves Saint Laurent to capture the fragrance of the orient. The perfume, ironically, was named Opium, after the drug imported to China by Western merchants and that helped end the traditional dynastic system of China.

This particular interweaving of places as contexts and commodities did not end with China but continued with exhibits and merchandising campaigns for France's ancien régime and for the designs of Yves Saint Laurent and Polo/Ralph Lauren.

Authenticity

A department store looks like a foreign country or a museum, and a museum looks like a department store. Los Angeles looks like Disneyland and Disneyland looks like Los Angeles. They are all landscapes of consumption and share the tensions of commodities. These places and modernity in general evoke strong normative reactions. They can be exciting, changing, and magical and also disorienting, shallow, and inauthentic. Underlying these normative positions are basic moral issues. These are discussed in the next chapter. Here I look at authenticity as a case study to demonstrate how these normative issues are reactions to the tensions in the landscapes of consumption and must be understood in terms of these tensions. Authenticity is also significant in its own right. Questioning authenticity seems to be a characteristic of contemporary intellectual life, and raising this issue with regard to place reveals the importance of place in our lives.

What do we mean by *inauthenticity? Inauthentic* and *authentic* are, in Cohen's word, "negotiated" terms.[99] The belief that places are inauthentic has its roots in our reactions to particular aspects of the tensions of commodities. Some, like Boorstin, express annoyance that culture changes and uproots things and moves them from context to context. Others, like the ordinary consumer or tourist, may be unaware that such juxtapositions and simulations are occurring. Still others, like Eco, may enjoy these simulations and view their acceleration as a genuine and authentic characteristic of modernity. The meanings of *authentic* and *inauthentic,* then, are not absolute but depend on real conditions and on our reactions to these conditions. Notice that I use the word *real* to describe conditions. Reality is more basic than authenticity. Authenticity may not be absolute, but it does not mean that all else is irrevocably relative. This point is worth considering in more detail.

The word *authentic* and its synonyms *genuine, veritable,* and *bona fide* assume that there is a reality, or at least the possibility of agreement about it. Certainly there can be claims that the real is also relative or even unknowable, as in the debates about realism and idealism, but these issues do not appear in the concerns of authenticity. Even the postmodernist Baudrillard assumes the existence of reality, despite his flamboyant attempts to obscure it. When he says that "Disneyland is there to conceal the fact that it is the 'real' country," or that "Disneyland is presented as imaginary in order to make us believe that the rest is real, when in fact all of Los Angeles and the America surrounding it are no longer real, but of the order of the hyperreal and of simulation," he nevertheless is making pronouncements that he hopes will be accepted in themselves as real and true.[100] He, like all radical skeptics, must assume a truth value to his statements. The assumption that truth and reality exist pervades his four stages toward simulation. First, he says, the image "is a reflection of a basic reality"; second, "it masks and perverts a basic reality"; third, "it masks the absence of a basic reality"; fourth, "it bears no relation to any reality whatever."[101] Note that reality is assumed in each of these stages.

In some cases, and especially in the consumer's world, there may no longer be a clear and stable picture of the things symbols refer to. Thus in Vreeland's world, the China that is exhibited is a China that refers back to the American consumer's vision of a China, molded by advertising. But this does not mean that it refers to nothing at all. Rather, this process of referring in the consumer's world can be thought of as a series of images and reflections. Consider a polygon constructed of mirrors that face inward. As soon as an image appears in front of one of the mirrors, it is reflected in the others, which also reflect the reflections. In one sense, this image of mirrors is much like the view of symbols held in the late Middle Ages and early Renaissance, in which symbols were thought to reflect each other according to the rules of *emulatio* and similitude.[102]

The difference between then and now is that there used to exist a hierarchy of symbols and a belief that these were designed by God and reflected his will. Even though in the present there may be no God, and the reflections may be extraordinarily complex, it is still possible to trace the image back to its first introduction and to see how the mirrors are arranged and constructed. Moreover, the analogy of mirrors does not pertain to all of the world but only the consumer's world. Beyond it are other systems of meaning, and the very fact that we can describe the consumer's world means that we can rise above it. True, the consumer's world is disorienting, but it does not have to be our only perspective, despite what postmodernists claim.

Although the relativity of authenticity and inauthenticity cannot, by itself, tear down truth and reality, can we reverse the problem and anchor authenticity to a definition of *reality?* Just such an attempt is made in several existential works, especially those by Martin Heidegger. Although these arguments are unsuccessful, they illustrate how relative the concept is and how it involves particularly modern problems.[103] For Heidegger, authenticity does not refer to what really exists but rather to the awareness of the conditions and responsibilities of human existence. Authenticity is the stripping away of appearances and the seeking of our essential selves. This awareness reveals what is truly authentic for human life and makes us responsible agents. Notice that authenticity is subjective. Its source is awareness. Indeed, our public roles are eschewed as shallow and suspect, diverting us from our true selves.

The reason authenticity is not anchored in this inward sense of truth is that it is not a philosophical absolute but a relative concept, which has changed historically. It is in fact the opposite of the sense of self accepted in the ancient Greek world (and also in non-Western cultures up to the sixteenth century), a sense defined in terms of performance of our public obligations and roles.[104] Unlike this older conception, the newer one claims that the public and social arenas draw us away from our true selves: modern society is alienating. If this characterization of society is correct, then we can understand why Heidegger seeks reality within the self.[105] But this very understanding transforms the search for an absolute into a historically relative reaction.

Authenticity raises important questions about the self and the world, about objective and subjective, but the attempt to ground authenticity within an absolute problem of human existence does not rid the concept of its relative components or prevent it from becoming a negotiated term, to use Cohen's phrase.[106] Authenticity is a problem in modern life because our culture creates a peculiar mismatch between the private (subjective) and the public (objective). The public world, the private world, and the sense of self are all in flux. While those arguing for an absolute meaning of *authenticity* call on the self as an anchor, this anchor is also drifting, because part of the modern condition is that self is fragmented and disoriented.[107] Commodities and their contexts constitute and accentuate these tensions, and large-scale places that are commodities or contexts for commodities do so even more.

Authenticity is a relative evaluation of modern life and place and stands as an indication of our reactions to modern life. Following Cohen, these reactions can be thought of as forming part of a continuum.[108] (The parts of the continuum are distinct only analytically, for each of us exhibits the range of reactions to even a single place.) It is useful to subdivide the continuum and consider each section as occupied by a certain personality.

The possible reactions to modern life depend in part on people's awareness of and adjustment to the tensions of commodities. People who are adjusted to the tensions or are simply not aware of them do not feel that the world is alienating or inauthentic. They do not see anything unusual in the juxtapositions of one place with another, in the attempt to make generic places appear specific, in the rapid turnover of contexts, or in the way place defines self and others. Nothing seems out of place.

People who are aware of the tensions may emphasize their positive characteristics. They may see modern reflexivity of meaning, for example, as leading to freedom and creativity, rather than to a burden of communication and a fragmentation of discourse. They may see a world of strangers as liberating them from stifling conventions rather than alienating them from their fellow human beings. They may see spatial segmentation and integration as balancing self-development and social engagement. And they may see the juxtapositions of contexts as stimulating, entertaining, and liberating—permitting self and group expression. To these people, many modern things and places can be out of place, but that is itself the modern context and is to be applauded.

Others dwell on the negative character of these tensions. They see danger in the fragmentation of self, in the juxtaposition and rapid turnover of contexts, and in the disorienting qualities of modern places, which are separated from the processes of their production and appear as finished contexts, which become images of other contexts. For these people, the modern landscape is alienating and inauthentic.[109]

Claims of inauthenticity are reactions to modern contradictions, which are intensified in the consumer's world. No wonder then, that the term *authenticity* is strongly relative. Places seem to evoke the term more often than advertising does. Reactions to advertising range from the positive—we find them clever, important, and entertaining—to the negative—we see them as deceitful and false. But they are not normally described as inauthentic. This could be due to the fact that for advertising we are willing to accept plan and authorship and thus to lay blame, while we are not willing to do this for the creation of place. Places do not usually have a single author. Places are also taken more seriously than ads. What, after all, can be more concrete and real than a place? In fact, one way of making things seem real is to turn them into places: for example, the past assumes a greater reality when it is circumscribed within the boundaries of an outdoor museum. When we see that places are not what they claim to be, our reaction could be strong indignation.

Arguing that *authenticity* is a relative term does not negate the possibility of objective reality or moral absolutes. The relativity of authenticity emerges most clearly in our reactions to the geography of the consumer's world. In the next chapter, I show how the relational geographical framework can become a basis for making moral claims that are not relative and that can be used to evaluate the morality of the consumer's world.

· III ·

Geography and Morality

· 8 ·
Place, Morality, and Consumption

Geography provides a foundation for morality, and geographical principles have moral implications. One consequence of this connection is that geographical place forms a basis for a common ground for divergent moral theories. Consider first the general issue of how geography and morality are connected.

Place as a Moral Concept

To make our homes, our workplaces, our world good or better is a general geographical concern. *Good* here means more than *pleasant* or *comfortable;* it means a place that helps us attain a good life anchored in the moral.[1] Regardless of our ranking of particular virtues and vices, we can agree that a community in which neighbors are spiteful and cruel, in which they lie, cheat, and abuse one another, a community full of violence and filth, in other words, a community in which the generally accepted virtues are absent simply cannot be a good place and prevents its inhabitants from living a good life. Indeed, such a place is a living hell. Good places, then, are seen as ones that uphold moral principles of one kind or another.

Geographers have not always held this simple point clearly in mind. Rather, their moral interests have focused on the more particular issues of the geographical distribution of values and customs, the injustices resulting from unequal geographical access and provision of goods and services, and the development of a land ethic and responsibility to nature.[2] Viewing place itself as a moral concept helps focus these geographical concerns about morality and, even more important, helps provide common ground for moral principles.

This second point follows from the fact that the categories of forces and perspectives on the intellectual map are permeated with moral implications, so that the contemporary intellectual fragmentation among these positions is reflected in a splintering of moral theories and perspectives. More specifically, ethical implications permeate forces and perspectives in three ways. First, forces are sources of ethical positions (except, of course, to those who believe in God as a distinct and separate source).[3] Some claim that ethics spring from the mind and the meaning we impart to the world; others, that ethical actions and meanings are formed at the most basic level by social relations; still others claim that moral behavior is an expression of natural or biological forces.

Second, ethical issues pervade forces because they offer the primary means to understand the moral consequences of our actions. To know if what we do is good or bad, we must know its effects, and how whatever we do will affect nature, meaning, and social relations. Ethical theories, like social theories, differ with regard to the emphasis they place on forces. Some assert that the morality of our actions should be weighed primarily in terms of their effects on nature; others, that morality concerns our duties and obligations to people in a social context; and yet others stress that morality has nothing to do with these overt effects but is purely a matter of intention or meaning.

Third, perspectives are also morally implicated. Morality constitutes one of the axes in the relational framework; and moral theories differ in the perspectives they assume from somewhere to nowhere. A moral theory such as emotive ethics is a lens close to somewhere, whereas a Kantian categorical imperative and a utilitarian calculus of the greatest good for the greatest number are lenses from nowhere.

The intellectual surface helps map out moral positions and points to geographical place as the primary means for providing common ground. Indeed, a principal reason for the fragmentation of the positions is that philosophers do not usually consider the role of geographical place in morality. As geography's role is clarified, a picture of the common ground will emerge, which can then be used to evaluate the morality of places of consumption.

The Good and the Moral

The good is related to the moral, and at times it is difficult to distinguish the two. Morality for Yi Fu Tuan provides a life that is "more committed, narrower, and more heroic than is a good life."[4] But sometimes the difference may be due more to the divergent moral perspectives that are used to evaluate an action than to some intrinsic distinction between good and moral. From one perspective, it might seem that denying ourselves luxuries and saving instead for our children's education is being a good and even moral parent. But from another perspective, this act may conflict with our moral responsibilities to our community and fellow human beings. The money we save for our children might do more good if it went directly to charity. Suppose we had enough money to become benefactors of our communities and endowed them with hospitals, schools, and libraries. We would then be acting as good or even moral citizens (though perhaps not as good parents). Still, according to some abstract moral precepts that seek a universal perspective (as with some forms of utilitarianism that strive for the greatest good for the greatest number) our actions may be too local or provincial to benefit those who most need them. From these abstract perspectives, our good and even moral acts may be less than optimal. I do not wish to argue that all differences between good and moral depend on how we define morality. Rather, I wish to point out that the good life is connected to morality, and morality in the modern world is not a homogeneous set of precepts but rather a group of fragmented viewpoints. One reason for this is that many moral theorists have neglected the role of place and context.

Consumption and Civility

The good life is not only linked to morality, it is also connected to consumption. As Tuan points out, the good life is one that requires a certain ease and material base, though it should be tempered by austerity.[5] Socialist societies, too, recognize that material comfort undergirds the good life. They wish to raise the standard of living of their members by allowing them more commodities. Unlike many capitalist societies, they have an avowed commitment to maintaining equality of access to commodities (often by owning them publicly) and perhaps a greater concern about the social contexts that give rise to them (though they may show less of a concern for the despoliation of the natural environment). The difficulty in all cultures is balancing comfort with austerity. Pâté de fois gras and gold-plated telephones are ready examples of excess. We should be uneasy about prosperity when it is unequally shared and comes from the exploitation of others.

How do we find a balance between a comfortable life and austerity? A general answer lies in becoming aware of the link between consumption and the three realms of forces. A realistic sense of austerity for our times, according to Tuan, could mean "not so much a diet of bread and water as openness to certain kinds of hard truths. How much of the world's comfort and splendor are we still able to enjoy in easy conscience if we have become more fully aware of their cost in the spoliation of nature and in the burden laid on people less fortunate than we."[6]

A historical connection between consumption and the good is found in the development of what has been termed the "civilizing process."[7] According to Norbert Elias, this process involves an increasing self-consciousness, an examination of our behavior from the perspective of others, and a tempering and muting of our emotions and drives by ideals of politeness, courtesy, and civility.[8] These are not in themselves virtues. Civility can become excessive and a veneer to cover up evil. But politeness, courtesy, restraint, and other forms of civility are necessary for real virtues to exist. Indeed, civility may result simply from increased social complexity. The larger and more differentiated our society becomes and more functions it contains, the greater the need for civility. "The web of actions must be organized more and more strictly and accurately, if each individual action is to fulfill its social function. [Thus] the individual is compelled to regulate his conduct in an increasingly differentiated, more even and more stable manner."[9] One result is an increase in politeness, courtesy, and civility.

Corresponding to this regulation of behavior is an increased reliance on symbolic systems that include material objects and functionally differentiated places. Objects and places—important devices for stabilizing and making visible our feelings and thoughts, for making a world—include a culture's cooking and dining utensils, its eating, sleeping places, its entertaining places, and its rules of conduct in public and private places. In assigning meanings of restraint and civility to these objects and places, the civilizing process is at once behavioral and material.

In a capitalist society, material objects become consumer goods, and our desire for these increases as the objects proliferate, because, according to Rosalind Williams, "the more behavior is watched over and passions curbed, the more consumer objects are complicated and the more they proliferate."[10]

In Western Europe at the end of the Middle Ages, members of the aristocracy became conscious of the civilizing process and its dependence on material objects and segmented places: they developed a complex courtly life, which was linked to the consumption of sumptuous material goods; their behavior was justified as a means to set standards and to underwrite art and culture. Consumption and courtly life were the means of advancing civilization. With the rise of capitalism, the power of the aristocracy waned and affluence spread beyond this group to large middle and upper classes, but the residual prestige of the aristocracy made it the trendsetter well into the nineteenth century. As prosperity increased and as it spread to more groups, it became important for these trendsetters to separate themselves from two contaminants: uncivilized elements of the community and their own unbridled excesses and decadence.

This concern to find a middle ground for civility was first and most clearly articulated in France. Here, even as the middle and upper classes increased in size, the prestige, if not the power, of the aristocracy continued to make it the bearer of culture. Its goal, which became the goal of all polite society, was a community where "manners are gentle, education broadly distributed, laws rationalized, and art and science cultivated."[11] This ideal spread throughout the affluent classes, so that "to many enlightened thinkers of the eighteenth century [including formidable intellectuals like Voltaire], it seemed self-evident that enlightened consumption—patronage of the arts, the vivacious conversation of the salons, collection of paintings and books—was a necessary means to the advancement of civilization."[12] Whether or not this reasoning was a rationalization, it was offered in earnest: it attempted to justify consumption by placing it beyond the interests of an individual and in the interests of society as a whole.

Although this view of civility and gentle society was not beyond dispute (Jean Jacques Rousseau decried consumption and luxury, unless all could share in it equally), it was popular. Consumption was seen as necessary to the development and maintenance of civilization, because the rich provided a market for cultural artifacts and thus preserved them. It also gave those who consumed a purpose beyond themselves: they could define their role socially and historically; they could become patrons of the arts and crafts; they could define and further refinement and taste. In short, they could be the bearers of the civilizing process. The legacy of these patrons—their grand houses and gardens and their vast art collections—to this day, form the backbone of Western high culture.[13]

With the decline of a social hierarchy, especially the aristocracy, the development of new styles becomes democratized and people look to each other for prestige. Our role as consumers becomes less a responsibility for the civilizing process than a satisfying of our appetites. In the age of popular culture, we can all become trendsetters, which makes it difficult to believe that our role as consumers upholds any other standard than our own self-interest. True, some influential people still create styles, and some forms of the mass media help set fashion. But their effect is muted, and their message changes constantly. Indeed, most advertising directs our attention to the fact that commodities are there for us and not for some abstract social responsibility. This isolation of consuming from social context and the concomitant isolation of the self have made each consumer responsible for deciding the purpose of consumption. In Williams's words:

> The implications of the consumer revolution extend far beyond economic statistics and technological innovations to intensely felt, deeply troubling conflicts in personal and social values. Before the nineteenth century, when only a tiny fraction of the population had any choice in this realm, consumption was dictated for most by natural scarcity and unquestioned social tradition. Where there is no freedom, there is no moral dilemma. But now, for the first time in history, many people in Western society have considerable choice in what to consume, how, and how much, and in addition have the leisure, education, and health to ponder these questions. The consumer revolution brought both the opportunity and the need to reassess values, but this reassessment has been incomplete and only partly conscious. While the unprecedented expansion of goods and time has obvious blessings, it has also brought a weight of remorse and guilt, craving and envy, anxiety and, above all, uneasy conscience, as we sense that we have too much, yet keep wanting more. . . . A part of us craves the rewards of "using up" the good things of life, while another part is aware of moving ever closer to the point of death, which will "sum up" our lives in a way that has nothing to do with transient pleasures.[14]

In the twentieth century, the connection between consumption and the civilizing process is no longer clear. Confusion abounds concerning the relation of consumption to the realms of social relations and also to the realms of nature and meaning. How does consumption affect these? Because of their fragmentation, theories do not provide a reliable answer. Nor does contemporary morality; because it relies on forces and perspectives to tell us the consequences of our actions, it too is fragmented. Just as place provides and integrates forces and perspectives, it can provide common ground for moral theories. I turn now to a consideration of the role of space and place in morality.

Geographical Foundations

How does geography form part of the foundations of morality? An approximation is provided by examining the way in which the questions of equality, freedom, and responsibility presuppose space and place. These presuppositions are evident even in the most abstract moral cases, and since much of contemporary moral theory employs a perspective from nowhere, that is where I begin. By reducing geographical issues to purely spatial ones, it is still possible for moral theories to overlook the geographical component. A lack of attention to geographical context, especially in abstract ethical theories, makes moral philosophy inaccessible and unrealistic, a point that has been noticed by several philosophers.[15] This does not necessarily imply that geography raises unique moral issues (though it might well do so) but rather that the conventional concerns of even the most abstract moral philosophy must attend to the geographical context.

Space, Equality, and Freedom

Equality and freedom have always been difficult to reconcile, as the extreme of either one interferes with the other. This is true for the narrow meaning of freedom that I consider here and that appears in the social science literature concerning choice, but it is even more true for the broader philosophical issue of free will. Geographers have paid more attention to equality, which has been conceived of in terms of geographical accessibility of goods, services, and resources, than they have to freedom. This emphasis can be explained by the fact that many geographical models are extensions of neoclassical economics. In these economic theories, freedom of choice is not a major problem, because it is assumed to exist within a larger competitive economic system; individual choices are thought to be constrained primarily by individual budgets. This means that freedom of choice, coupled with different incomes, leads inevitably to unequal consumption. But one area of neoclassical economics where differences of consumption should not exist even in theory is in the provision of public goods and services.

A pure public good is one that can be provided to everybody equally, such as national defense or the legal system.[16] Geographers have argued that a test of the principle of equality is equal spatial access to the provision of public goods and services. Spatial accessibility is not always translatable into real access—poor people can be near inner-city hospitals that will not treat them because they do not have medical insurance coverage—but it at least provides a rough index and is often a necessary condition to accessibility.[17] A theoretical discussion along these lines is the tack taken here, except that it does not rely on a particular economic system; the argument considers only the relation between space and the provision of public goods.

The problem of geographical equality begins with the recognition that natural and social resources are distributed unevenly over the earth and that not every place is equally accessible to all resources. This unevenness is a fact of nature, and is a focus of one of geography's most enduring interests—the areal variation of phenomena over the earth and the differentiation of this variety in specific places.[18] This differentiation leads to uneven patterns of human settlement. Wealthy nations have the power to reduce and even to strive to eliminate spatial inequalities of the goods and services they provide: they can attempt to distribute health care, education, and legal protection equally to all of their citizens regardless of where they live. But can even the wealthiest and most powerful nation succeed? Can there be complete geographical equality of access to public goods?

A thought experiment can illustrate how difficult this can be even in theory and how the attempt itself introduces other moral problems. To simplify things, I employ the well-known geographical device of ignoring the geographical reality of areal differentiation by reducing the natural and social environment to a uniform, isotropic plane. This means that people are distributed uniformly over a flat, undifferentiated surface, that roads and resources exist everywhere, and that the content of one location is exactly like that of another. Justice is defined in terms of equality of access to public goods and services, of which there is a sufficient quantity, so that the demand for these items does not lessen their supply. Theoretically, such public goods are offered by governments and take the form of legal protection, military defense, and more locally, the services of institutions such as public schools and police and fire departments. Thus if we need to avail ourselves of the courts, our use of this service does not prevent someone else from using it.

Let us suppose that these goods are public and that their supply is virtually inexhaustible. Yet courts, schools, and police and fire protection are provided by people and equipment that are virtually impossible to spread to every location equally. These services have to be geographically localized. Fire protection requires fire trucks, which need to be housed and serviced, and firemen, who must be in the firehouse and at the ready. The police need a station house, with jail cells, interrogation rooms, and offices. And education takes place in school buildings. Someone's house will be farther than someone else's house from one or another of these services, and this fact alone creates inequality in accessibility. (Ironically, being too close to the source of a public good may also create problems—noise and chaos in the neighborhood of a fire station, a police station, a school, and even a public park.) If we push this thought experiment to the limit, it is possible to imagine that, with ever more resources and effort, geographical inequalities can be reduced to the point where they no longer matter. Whether this end point is indeed an attractive one again depends on how highly we value equality of access. In any case, this part of the experiment reveals the importance of spatial considerations in a discussion of equality and the enormous investment that must be made in geographical infrastructure.

The experiment is geographical only in a rarefied sense. Virtually everything of geographical interest has been eliminated except the distribution of public goods. And, as we shall see, the experiment's stringent assumptions severely constrains our freedom of choice. Before considering directly the question of freedom though, we need to carry the thought experiment one step farther in order to illuminate another side of the geography of equal accessibility—the geography of responsibility, which of course implies restraint on freedom.

The experiment assumes that, in any isotropic plane, we have geographically uniform access to goods and services, such as schools and fire and police protection, but the experiment does not consider the geographical extent or the number of places that provide these services. Thus equality of access means that the responsibility for providing these goods must be assigned to specific places, and this assignment of responsibility constrains freedom and interferes with the equality it was intended to support.

Given a large number of people distributed uniformly over a vast and geographically isotropic plane, many schools, police and fire stations, courthouses, and so on may be required to serve them all, and these services must be identical everywhere. Suppose that a child on one end of the isotropic plane, a budding geographer who enjoys travel, wishes to be bussed an indefinite distance to attend school at the other end. Granting permission to this child and to others to attend a distant school would affect the enrollments in each school, the accessibility of students to teachers and instructional resources, and thus the geographical equality of public goods. True, the assumption of equality may eliminate any reason for the students to decide to travel to a more distant school (even a desire for travel could be met in an isotropic plane by going around the school in circles); nevertheless the experiment does point to the fact that the relation between people and sources must be thought of territorially. That is, a school draws its students from within a territory and is responsible for educating only those students, and a police department cannot choose its clients but must protect anyone in its jurisdiction, citizen and visitor alike. People cannot choose which police department or which police officer will protect them; nor can students choose which school to attend.

Assignment of responsibility leads to a territorial definition of social relations.[19] This means that we are assigned to a particular group or community simply on the basis of our location within a territory. In this respect, we are treated as all others are within that territory, and the providers of goods and services are responsible only to those within the territory. Thus even in the simplest of cases, territory becomes an indispensable means of defining responsibility in a system of equality. Territoriality's indispensability depends on the advantages inherent in territorial control (which are explained by the theory of territoriality). Among the simplest and most benign advantage is that territoriality provides a clear and unambiguous means of classification. If we are classified according to attributes other than our location—perhaps by our age, income, or sex—we would receive goods or services according to these attributes, which would not be equal treatment. In this regard, a territorial definition helps preserve equality, because it treats everyone in an area in the same way. This is in fact one of the major reasons why public goods are provided territorially.[20] Yet, territorial control itself can set in motion its own inequalities.

In the isotropic plane, the system works only if uniformity is maintained everywhere, and this means that people and resources cannot shift position without some accounting procedure that weighs the advantages and disadvantages of these shifts and makes countermoves to make everything equal once again. Such an accounting procedure would include a hierarchy of territorial responsibilities, so that—as in the Catholic church with its parishes, dioceses, and archdioceses, or as in the American system of municipal, county, and federal governments—there is a larger unit that oversees the activities of smaller units.

This hierarchy of units and responsibilities makes the degree of knowledge and control different for those administering the larger units than for those administering the smaller units. Unless everyone shared equally in the administration of all levels (which would deny the need to administer in the first place) a hierarchical territorial structure that was put in place to assure equality can itself create the potential for unequal knowledge, power, and responsibility. This potential for inequality is not a minor wrinkle in the landscape of equality: consider how corrupting territorial authority became in the Catholic church, an institution avowedly focused on pure and otherworldly concerns. The fact that there are many church canons prohibiting priests and bishops from leaving their parishes and diocese (called translations) to accept positions in larger ones reveals how control over a specific territory in a hierarchy was often regarded as an end in itself.[21] Being an archbishop of a larger territory gave one more power and prestige. As Bishop Hosius cynically observed, "the cause [of translations was] well understood; for it had never been seen that a bishop left a large bishopric to take a lesser one."[22]

While a territorial hierarchy can itself be a source of problems, complications can also arise at the lowest territorial levels. We assumed that a police department is responsible only for crimes within its jurisdiction. Suppose that, in spite of our efforts to equalize everything, simple random events suddenly increase the crime rate in one territory. Where does the responsibility to fight these crimes rest? A territorial hierarchy, with an accounting procedure, could allocate resources for such an emergency, but this hierarchy could also foster a parochial outlook that would hinder generosity. Responsibilities would be limited to one place unless a larger territorial authority tells us otherwise, which in and of itself may not be a bad thing—after all we cannot all be responsible for everything everywhere. But is it a good idea to make location a major factor in molding our sense of responsibility? A territorial definition of social relations seems arbitrary. Should it be a base for our concerns for others? Even in the isotropic plane, we may be affected more by what occurs outside our territory than by what goes on within it. For example, if a community wishes to make its distribution of goods more equitable by moving some of its services to the periphery of its boundaries, the move could create negative spillovers or externalities in the form of noise or unsightly nuisances for those living near the boundary in adjacent communities. A territorial definition of responsibility may place external effects low on a list of priorities. This problem also raises the static nature of boundaries and the fact that they need to be continually readjusted to minimize mismatches between responsibility and effects.

Even though territoriality engenders difficulties, it is often the most effective means of defining responsibility. Its advantages in classification make it the most practical solution, even if the geographical conditions of the experiment are changed, so that there is equal access, plus freedom of movement and uneven distribution.[23] In a system in which resources and people are distributed unevenly and where things move and interact at various rates, the theoretically most equitable solution might be to have peripatetic teachers, officers of the court, police and firemen, and other purveyors of public goods assigned to specific individuals in this mobile population. But since this would be impossible, a realistic balance between efficiency and equity would be to have goods and services distributed territorially and to try for a match between supply and demand within that unit. This is in fact the method used in the real world, where resources and people are unevenly distributed and geographically mobile.

The advantages of territoriality for public goods are the same for nonpublic goods targeted to specific people. For example, suppose a major appliance manufacturer provides all customers with a one-year warranty that includes home repairs. If any customer within a year of purchase has difficulty with an appliance, the manufacturer will send a service person to the customer's home, free of charge. Obviously, the distribution of customers could be extensive and geographically dispersed, and customers may move residences. Still, the company may well find that a territorial assignment of repairmen may be the most effective way to serve its customers. Thus even if you purchased your appliance at place *x*, if you reside in place *y*, you will be serviced by repairmen in charge of place *y*.

These hypothetical problems illustrate how intertwined spatial and moral issues are and how inescapable geographical considerations are. Thus geographical equality and freedom often move in opposite directions, and territoriality, though of help in assigning responsibility, may introduce its own complications through mismatches, spillovers, and a focus on the local or provincial. This last issue is taken up next, because the local or provincial is important in moral discourses about community, responsibility, and morality. The question of our moral purpose in life is linked to how we define our responsibilities, including being a member of a community. If our community is local, then so are our obligations. But if our community is unbounded, how do we define our purpose? Without a clear sense of purpose in life to provide an anchor, moral discourse drifts between the extremes of altruism and absolute equality, on the one hand, and self-interest and unbounded freedom, on the other.

Propinquity, Community, and Responsibility

We are social beings; we define our purposes and ourselves in terms of our relationships to others in communities. Behind the idea of community lies the question of geographical propinquity.[24] A popular claim (reflected in the nostalgic and local strain of modernity mentioned in chapter 2) is that the real or authentic community, which is assumed to be a desirable state, must be local. A local community is presumed to occur when the spatial manifestations of a particular system of production, consumption, and other social relations overlap and are enclosed within a single place. For the people living there, the place becomes their world. From this perspective, the more our social relations are geographically diluted and dispersed, the more they deviate from genuine local community. Deviation can also arise if some social relations are artificially imposed by territoriality. A territorial definition of social relations, such as the U.S. practice of allowing people to become political members of the communities they reside in, assigns strangers to the same community, even though they have little else but propinquity in common.

Modern social and technological changes are thought to have weakened local communities by allowing the web of human interactions to extend every which way in space. Thus communities that transcend the local, such as world religions and international scholarly communities, are somehow unusual or incomplete; similarly, modern communities that are largely territorially defined, such as a modern neighborhood, city, county, or state, are also incomplete or artificial. This conception of community and its transition is reflected in such generally accepted distinctions as *gesellschaft* versus *gemeinschaft* and organic versus mechanical solidarity.[25]

The conception of local community as an ideal is found in utopian literature and in the nostalgia about the way society used to work. It is also reflected in older political theories of representation that hold that a single person can represent to an assembly the interests of a town, city, or county, regardless of its population, because such local places were organic communities whose interests were undivided and self-evident. Representation by area is still a common practice, but it is not often expected that the interests of the area will be uniform, and proportional representation is usually thought of as more just. Still, the retention of representation by area, as in the U.S. Senate (where each state, regardless of population, is represented by two senators), reveals a lingering sense that local areas (or states) may be, to some extent, genuine communities of interest.

People who wish to reestablish these communities may have doubts about the effectiveness of constructing them through territorial controls; although a territorial definition of social relations can lead to genuine feelings for one another, in our dynamic world, such territories alone cannot circumscribe many of the activities that define our primary obligations and responsibilities to others.

I do not wish to argue that this conception of a local community ought to be upheld or that such places were more likely to occur in the past than they are now, though I do think historical and ethnographic evidence does support this last position. Rather, we need to see why proponents of such a meaning of community claim it is a good thing, and why others do not. Several arguments can be offered in favor of local community.

Responsibilities forge our purpose in life. If our responsibilities to our kin, our political community, our work, and our home depend on local encounters, these responsibilities are reinforced by shared experiences, which immediately and tangibly reveal how our obligations in one sphere affect those in other spheres. These encounters help us understand the consequences of our actions. Local community can take advantage of the weightiness and reality that come from a web of personal contact. All of the electronic modes of modern telecommunications put together do not equal the complexity and weight of a personal encounter. (Modern telecommunications fragments the experience of communicating and thus narrows our sense of obligation.)

Face-to-face encounters not only add weight to interactions, they take time, and some argue that time is important in developing a sense of responsibility and an understanding of the consequences of our actions. Local communities, the argument continues, provide the best, though perhaps not the only, means by which we develop a clearly defined role and a lasting purpose. Understanding our roles and purpose makes it easy for us to evaluate our actions. Further, local community equates responsibility with place and encourages the establishment of roots. (Such communities, though, would still need territoriality to bound them, and such a boundary could be used to constrain new activities and, thereby, exercise a territorial definition of social relations.)

This picture of local community is lent further support when embedded in the Darwinian view of evolution. This view maintains that geographical dispersion of diverse life forms into different and isolated geographical contexts or niches has the benefit, through the intermixture of life and habitat, of increasing each species' chances of survival and of increasing the abundance and variety of life. The implication of local community is that it is relatively isolated and forms a self-sufficient ecological niche. In Ellen Semple's words, "the divergent types of men and societies develop[ing] in segregated regions are an echo of the formation of new species under conditions of isolation which is now generally acknowledged by biological science."[26] This picture of local community points to the additional virtue of providing a multitude of social experiments that increase the chances of human variety and survival. This supposition is embedded in American thought, where, according to J. Nicholas Entrikin, "the regional diversity of ways of life was considered to be an important element in the maintenance of democratic institutions, and thus an aspect of life to be valued and maintained. . . . [The nation] derive[s] strength and unity from diversity."[27] Those who believe in local community see the modern dynamic world system as a threat to the viability of these places.

Virtually every argument in support of local community can be turned around and used against it. It could be said that local communities smother their members in traditional obligations; that they stifle change; and though they foster responsibility and purpose, they make these moral precepts so narrow and provincial that everyone who is not a member is a stranger to whom the community has few obligations. Rootedness and locality become more virtuous than mobility and cosmopolitanism. In short, local community interferes with the dynamism of modern life and with the expansion of moral responsibility to include all human beings and even all of nature.

The two viewpoints are more dialectical than antithetical; they share some common ground. They both suppose that communities were once more geographically coherent and tightly knit. The problem concerns the suitability of this community for a modern world and the kind of morality that would emerge with the weakening or even the absence of the local community. Its supporters argue that, even though this community may have its limitations, it remains to them unclear how a moral point of view can be nurtured in the absence of a genuine local community. Can we think of the whole world as such a community? If so, then how can we be truly responsible for everybody and everything? Is a global morality possible? What would it be like?

If the terminology shifts from geographical to philosophical, these problems also shift to the general concerns about the degree to which ethics should be concrete or abstract. Can moral theory be successful as it moves from somewhere to nowhere?

Moral Philosophy

An important issue in moral philosophy is whether our moral precepts are relative or absolute. Are they based on particular roles, obligations, and contexts, or do they apply universally? Is morality embedded in a somewhere, or can it be formulated from nowhere? The alternatives are the ends of a continuum of answers that contemporary moral theories provide. But these alternatives also point to a historical progression in the attractiveness of the answers. Abstract moral theories have gained increasing attention. This progression in interest is closely linked to changes that have moved communities from a parochial to a global context. This progression also reveals how closely linked abstract moral theory is to geography.

According to MacIntyre, this progression in Western philosophy went through three stages.[28] The first he calls the heroic age, referring to a form of culture that produced the great sagas of Ireland (the Ulster cycle), Iceland (the Icelandic sagas), and especially Homeric Greece (the *Iliad* and the *Odyssey*). These tales were infused with moral conflict, and the moral precepts were not universal but rather specific to each culture and to individual roles. In these cultures, "morality and social structure are in fact one and the same," and in each of these tales, "the basic values of society were given, predetermined and so were a man's place in the society and the privileges and duties that followed from his status."[29] The heroic age did not permit people to see themselves outside of their context. A position of nowhere made no sense, because a person was real only in terms of his duties and obligations.

> The self of the heroic age lack[ed] precisely the characteristic . . . that some modern moral philosophers take to be an essential characteristic of human

> selfhood: the capacity to detach oneself from any particular standpoint or point of view, to step backwards, as it were, and view and judge the standpoint or point of view from the outside. A man who tried to withdraw himself from his given position in heroic society would be engaged in the enterprise of trying to make himself disappear.[30]

The moral precepts or virtues of heroic societies facilitated particular roles. They included bravery, cunning, a sense of humor, friendship, and loyalty. They did not include humility, charity, or universal love.

The second stage of philosophical development is the classical period, characterized by a movement away from somewhere. The classical view attempted to pose moral issues abstractly, lifting each individual above her or his role in the family and work. But the individual was never conceived of without context. A person was still a social being and bound to a particular community—in the case of Greece, to the city-state. "The common Athenian assumption then [was] that the virtues [had] their place within the social context of the city-state."[31] This does not mean that there was agreement about what virtues were. Some, like courage, might have been held over from the heroic period, but others, such as justice, self-restraint, wisdom, and citizenship, would be new, and different philosophers might have assigned each a different importance. Still, for Greek philosophers the virtues were conceived of within the context of the city-state and remained a political concept. For Plato, "the virtuous man [was] inseparable from his [role as a] virtuous citizen."[32] Aristotle held a similar position; from his perspective, according to Richard Norman,

> One's moral education consists in being told in particular situations that one's behavior is appropriate or inappropriate. . . . In this way one builds up an intuitive sense of when, and to what extent, anger is appropriate. . . . Now, a crucial point is that the matters about which one acquires this sense are very much matters of degree. This is why one cannot formulate any precise rules about them. . . . This is why Aristotle thinks it is impossible and unnecessary to say any more than that the mean is "at the right time, to the right degree, on the right grounds, and towards the right person" and that the standard is fixed by [those with wisdom]. The knowledge which enables us to understand this is acquired not by learning theoretical principles, but by moral training, by being properly brought up in a morally civilized community.[33]

Having morality stem from one's day-to-day obligations makes one suspicious of categorical injunctions that have no context. Thus to Aristotle, generosity in the abstract is not a virtue, because a virtuous person "will refrain from giving to anybody and everybody, that he may have something to give to the right people, at the right time, and where it is noble to do so."[34]

To Plato and Aristotle and their followers, the virtues were compatible with, and reinforced, each other. The good life was "itself single and unitary, compounded of a hierarchy of goods," and conflict among the virtues would be "simply the result either of flaws of character in individuals or of unintelligent political arrangements."[35]

In the present period (the third stage), the social fabric has dissolved, and the immediate social roots of duties and responsibilities are unclear, so that morals are conceived of in abstract and universal terms. This has led to attempts to derive all moral precepts from a single set of axioms (which reveal that logic, at least, is still held to be universal). According to David Miller, "a powerful thrust in . . . ethical theories [is a move toward] universalism: namely, the view that the subject matter of ethics is persons considered merely as such, independent of all local connections and relations."[36] From this perspective, all "particular injunctions should be derived from universal, rationally grounded principles," and many universalist positions make "room for individuals' particular duties, responsibilities and rights—the duties of parents, colleagues, and so on."[37] But these are derived from universal principles and would not be thought of as fundamental moral commitments.

Perhaps the most famous attempt at a universal ethic is Immanuel Kant's categorical imperative, which offers a single and universal statement of what constitutes a truly ethical or moral injunction. Kant says that one should "act only on that maxim whereby you can at the same time will that it should become a universal law."[38] The categorical imperative and other attempts, such as utilitarianism, to universalize ethics are all problematical. Either they are internally inconsistent, or they do not incorporate the moral precepts that others hold to be important and thus do not make allowances for real and complex choices.

An indication of how far modern culture is from a consensus about the content of morality can be gleaned by listing the virtues that contemporary society would want to include in a universal moral system. The virtues would be far more varied and disjointed than in an earlier time. Perhaps the best examples of how ad hoc they appear is Benjamin Franklin's catalogue of thirteen virtues in the form of maxims requiring obedience.

> These names of virtues, with their precepts, [are]: (1) Temperance.—Eat not to dullness; drink not to elevation. (2) Silence.—Speak not but what may benefit others or yourself; avoid trifling conversation. (3) Order.—Let all your things have their places; but each part of your business have its time. (4) Resolution.—Resolve to perform what you ought; perform without fail what you resolve. (5) Frugality.—Make no expense but to do good to others or yourself; i.e., waste nothing. (6) Industry.—Lose no time; be always employ'd in something useful; cut off all unnecessary actions. (7) Sin-

cerity.—Use no hurtful deceit; think innocently and justly, and, if you speak, speak accordingly. (8) Justice.—Wrong none by doing injuries, or omiting the benefits that are your duty. (9) Moderation.—Avoid extremes; forbear resenting injuries so much as you think they deserve. (10) Cleanliness.—Tolerate no uncleanliness in body, clothes, or habitation. (11) Tranquility.—Be not disturbed at trifles, or at accidents common or unavoidable. (12) Chastity.—Rarely use venery but for health or offspring, never to dulness, weakness, or the injury of your own or another's peace or reputation. (13) Humility.—Imitate Jesus and Socrates.[39]

If we add to Franklin's aphorisms the Ten Commandments, the seven deadly sins, and other maxims and aphorisms that either are vestiges of older moral principles or have developed to fit recent conditions in our culture (such as those, including Franklin's, that intone the virtues of thrift, hard work, and economic success), we see that they add up to a bewildering, fragmented, and often contradictory array. This fragmentation reflects what many social critics claim is the current fragmentation of social relations and the replacement of a conception of humans as social beings with a conception of humans as isolated individuals, who define their own sense of self, and who no longer possess life projects and purposes but, rather, have multiple roles and lifestyles.

Many moral theorists deny the possibility of a universal moral perspective and even claim that morality is really emotivism, a position (whose foremost proponent is C. L. Stevenson) that holds simply (following MacIntyre) that any moral statement is an expression of preference couched in lofty terms. Thus if I say, "This is the moral thing to do," what I really mean, according to emotivism, is, "I like this."[40] Emotivism is the denial of a universal morality and a belief that morals are contextual (i.e., are mores), yet it also recognizes that the modern world contains few clear contexts that nurture morality, and so, in desperation, it turns to the contexts defined by the isolated individual as providing the basis of moral meaning.

According to MacIntyre, with the breakdown of the social contexts that made purposes and projects clear, we now have isolated individuals who must invent their own morality because "modern moral utterance and practice can only be understood as a series of fragmented survivals from an older past. . . . [This makes deontology] . . . the ghost of conceptions of divine law" and teleology the ghost of older views about roles and community.[41] Yet, others like Tuan claim that, even though we have not developed abstract moral precepts for a global community, we nevertheless have made moral progress.[42] Becoming more universal is in itself a good thing.

This discussion of geography and morals leads to the following possibilities. At one extreme is the possibility of a universal moral view that pertains to a dynamic and global community in which the geography of our responsibilities extends to the far reaches of the earth. At the other extreme is the possibility that real purpose and moral consensus have been severely weakened by the diminished importance of the local community. Unmooring morality from the local community allows the moral to drift into fantastic abstractions, such as pure altruism, or dissolve into emotivism.

Of critical importance for both geography and morality is the fact that, at either extreme and at all points in between, space helps constitute morality. Whether in the local or global arena, our behavior affects others through causal links in space and thus helps constitute moral purpose and responsibility. Indeed, this role of space is even more important to bear in mind now since the local community is dwindling; we no longer have the face-to-face contacts that make tangible the spatial basis of responsibility. Geography, then, is a necessary condition for understanding duties and responsibilities. How these are characterized may depend on which moral theory is used, but all moral theories possess a geographical basis common to all humans. This geographical ground for agreement must be mapped further and then applied to the places of consumption.

Geographical Grounds for Agreement

The geographical common ground is based on the precept that moral agents are responsible for their actions. Most moral theories, though, do not emphasize responsibility, because they assume that the consequences of our actions, and hence our responsibilities for them and to others, are readily determined due to the fact that agents are similar and have limited and circumscribed powers and effects. These conditions may have been true of previous times, when responsibility was left to the most powerful—who was God—and when, in Hans Jonas's words, "the good and evil about which action had to care lay close to the act, either in the praxis itself or in its immediate reach, and were not matters for remote planning. This proximity of end pertained to time as well as space."[43] In the past, knowledge about good and evil was not specialized but was accessible to everyone. These assumptions no longer hold. The human condition is not fixed, "given once and for all"; no longer is its "range of action, and therefore responsibility . . . narrowly circumscribed."[44]

What has changed is our enormous and almost godlike power to transform every part of the world, be it the social, the intellectual, or the natural. This means that our behavior and its consequences are not static or limited in either time or space. This power places the responsibility for the consequences of our actions in the forefront of moral concerns. It should even lead us to a new moral imperative: "act so that the effects of your action are compatible with the permanence of genuine human life."[45] This concern with responsibility does not mean that we can actually know what the consequences will be. But it is immoral to act unless we have some assurance that we will not make things worse. This need not necessarily lead to a freezing of action.

The consequences of our actions cannot be determined without knowing their effects in space (and time): we must know their geography. This makes responsibility inextricably spatial, or geographical. Geography provides the same basic role in sustaining purpose. We cannot know if we are conducting ourselves properly without determining the geographical consequences of our actions in space. Geography, then, is necessary to moral responsibility and purpose.

The necessity to morality of a geographical view can be put even more forcefully by focusing on the connection of place to forces and perspectives. Our actions have consequences on nature, meaning, and social relations, and yet our theories are too fragmented to help us trace and weigh our effects on these realms. Where theory fails, place can help. Examining our effects in the context of place will force us to consider the spatial relations among all of the realms and at various scales. That is because place is central to them all.[46] As for perspectives, geographical place possesses an inside and an outside and thus allows us to view things from somewhere to somewhere else or to nowhere as a limiting case. It is through the somewhere, moreover, that the realm of agency becomes visible. Thus the somewhere reminds us that we not only produce certain effects but that we also have choices and can do otherwise as free agents. Thinking geographically allows us to view ourselves in place—and as free, and thus morally responsible, agents—and combines both the literal and figurative meaning of putting ourselves in the place of others. Indeed, no other way of thinking provides these advantages, as well as ensuring that consideration will be given to all three realms.

Responsible action, then, requires geographical thought. But this geographical basis of responsibility does not, by itself, entail other more specific moral imperatives, because further moral evaluations of actions depend on the priorities given to the interconnections among the forces and perspectives. These evaluations immediately raise the issue of the relativity of social theories and moral perspectives. Consider the comparatively simple case of the effects of purchasing commodities made of ivory. We know that the use of ivory endangers the African elephant. But we also know that poor people have built a livelihood around its sale. If we decide to boycott ivory, we will be placing greater importance on preserving an element of the natural environment than on preserving a particular social relation (see figure 29). The problem is complicated further by the fact that a more persistent threat to the elephant stems from population pressure and the agricultural practices of post-colonial African societies, which threaten to destroy the elephants' natural habitat. Focusing on ivory may obscure the wider issue of the relation between demographic and economic expansion, on the one hand, and the preservation of natural habitats, on the other.

The same dilemma is encountered in the consumption of beef raised in the clearcut areas of the rain forests of the Amazon River basin. If we decide to refrain from eating this meat, we will be placing a higher value on preserving a natural ecosystem than on assisting a particular form of economic development (though our choice here could be simpler than in the elephant example if Susanna Hecht and Alexander Cockburn are correct—that this is not an issue of economic development but of tax write-offs).[47]

In the two preceding cases, we make decisions based on the moral significance we attach to the consequences of our actions. Others may agree about these consequences but evaluate them differently. Regardless of these differences, without a geographical understanding of Africa and its animals, habitats, and humans, or of Brazil and its rain forests, economic policies, and culture, or of the connection of both places to the rest of the globe—understanding that involves the relations between meaning, nature, and social relations at various geographical scales—we would not be able to establish the connections upon which moral judgments can be made. This role of geography as a necessary condition in evaluating responsibility and purpose will serve, then, as the geographical ground upon which to examine the consumer's world.

The Consumer's Paradise

The commodities and contexts of mass consumption are immensely attractive the world over. Most people hope to raise their standards of living, and it has become an American ideal to let virtually nothing interfere with our right to afford the commodities of a consumer society. To many people, the world of consumption is equated with the good life.

Places of consumption and advertisements imply that what they sell leads to the good life. Many commodities make this claim overtly, as in advertisements that have us believe that life does not "get any better than this." The contexts that consumption creates have even been described as "utopias," as "paradises," even as "heavens on earth." The enormous attraction of mass consumption, though, is not based entirely on the success of advertising. The appeal of consuming has much to do with the way it combines forces and perspectives from around the globe, by placing the consumer in the center of a world of prosperity and ease. As we have seen, prosperity and the good life are interrelated, and there is something about the kind of prosperity that the consumer's world offers through its integration of nature, meaning, and social relations that should be taken seriously. It will shed light on our own culture's image of the good life and provide a means of comparing the good life to purpose and responsibility.

Let us consider the way in which the consumer's world can be thought of as a paradise, which means that we examine consumption on its own ground, as though its effects on the other realms were neutral or benign—as though its backstage has nothing to hide. It also means that we consider what people have in mind when they use the term *paradise*.

In Western thought, paradise is often linked to heaven. Although it has been portrayed in a number of ways over the centuries, most of these portrayals share important attributes.[48] One of these is that paradise, or heaven, is a place (although its location is often unclear). Another is that the inhabitants of paradise are sentient beings who experience pleasure. Another is that paradise perfectly and harmoniously integrates features from each of the realms. The natural setting might vary from clouds to gardens, but it is always comfortable and serene. Social relations might involve only God and his "children," or they might include a hierarchy of saints, angels, and the redeemed, including our ancestors. Meaning pervades paradise; its inhabitants enjoy and marvel at God's blessings. Meaning, nature, and social relations are brought together in perfect harmony, because paradise is a perfect place, a divine creation.

These perfections are often presented in concrete terms, perhaps to make paradise attractive to those of little faith. The Bible describes Eden as a perfect garden that provided Adam and Eve with every material comfort. Heaven is portrayed as a land of milk and honey. The Koran describes an even more elaborate paradise, containing fountains and gardens planted with all forms of fruits, surrounded by rivers of water, milk, and wine, wherein men and women, released from all toil and affliction, recline on raised couches. Absent from most paradises is labor, unless one is charged with God's work. (The principal exception is Marx's nonreligious workers' paradise.) Yet their inhabitants are provided a cornucopia of food and drink and an environment that gratifies every sense. Pleasure is had through the enjoyment of others and through the beauty of the scenery. The pleasures of paradise are a reward for a good, moral, purposeful, and often painful life. And of course, it is designed by God.

Suppose, though, we are no longer clear about just who the good or moral people are or about the kind of life that contains purpose; suppose we doubt the existence of God. Even a godless paradise would have to integrate successfully earthly forces and perspectives. And such a place would have virtually open admission, because in a society where humans determine meaning, and conflicting moral theories abound, virtually any standards could appear arbitrary or even prejudicial. Such a paradise becomes simply a pleasant, harmonious place, with each of us at the center, a place of immediate gratification. This looks very much like the kind of place consumption claims to have created. The consumer's world puts forward an earthly paradise, which does not possess divine purpose and for which admission does not depend on any criterion save financial success.

On the surface, the world of consumption glitters with excitement and change. It allows anyone with money to enter, and even those who cannot enter can window-shop. It treats us all with the same care. It lifts us from provincialism to a global community. It ties together things far and wide. It provides a means of expressing ourselves publicly. It makes us feel as though we are at the center of the world. A global community, the empowerment to create contexts, prosperity, and ease are all extremely attractive and seem within our grasp. People in this world do not want to give it up, and hundreds of millions of others aspire to join them.

But not everything in the consumer's world is smooth, even on the surface. The threads that it interweaves threaten to unravel. This is a world without constraints and without responsibility. It makes each of us the arbiter of what is important and of how much to consume. How do we choose one thing over another when there are no clear obligations and responsibilities and when there is no necessity? How do we form social relations and define ourselves when we have no particular projects or tasks? The unrestricted freedoms of the consumer's world could also create a weightless and disorienting world.

With greater affluence, there is the risk that we will no longer consume things as much as metabolize them.[49] There would be little time to linger over things, and time is important for enjoyment. Metabolizing the environment could dissolve objects of our world into stimuli, and ourselves into receptors. It could annihilate the difference between subjective and objective—could even annihilate consciousness itself. Then we would have no meaning to impart to objects. Commodities would lose their attraction, and perhaps consumption would simply run out of steam. Simply accentuating many of the tensions in the consumer's world leads to these consequences.

But so far, consumption has not run out of steam. And while the consumer's world can offer many examples of bored, disoriented people leading meaningless and self-indulgent lives, there are also many other people who, though they have extraordinary wealth, exercise restraint. Indeed, some measure of restraint is imposed on us by our own biology and mortality. We cannot stand constant stimulation. We have only so much time in the day, and our days are numbered. But restraint can also be imposed from within. Many with wealth may not indulge in conspicuous consumption and in fact lead austere lives. They may also enjoy the objects they consume and the contexts they create, so that they need not change contexts again and again. People can find purpose and meaning in the care they give to their possessions. And while this purpose and meaning might seem superficial by some standards, it might be weightier than the alternatives that many people find in their tedious and trivialized jobs.

On the surface, then, the consumer's world is not perfect, but neither would paradise be perfect without a divine order and a clear policy of admission. The drawbacks to the consumer's world become more serious when we go behind the surface to the backstage that supports it—to forces and perspectives and to the role of geography in defining purpose and responsibility.

As for the forces, we already know that there is a difference between the way they look from inside the world of consumption and the way they look from outside. Even a paradise must have support, and the consumer's world both draws on and affects meaning, nature, and social relations; but this paradise is so structured that it disguises the threads from these realms and weaves its own contexts, which appear to arise out of nowhere. This means that, from the perspective of the consumer's world, it is virtually impossible to understand the real effects we have on the realms. A shop that sells Colombian coffee does not reveal the social structure that produces the coffee, the economic impact of coffee production on the Colombian economy, or the way coffee growing affects the Colombian environment. Instead, Colombian coffee is advertised as though it were picked by fictitious planters, like Juan Valdez, in a coffee-producing paradise.

That these connections are not disclosed and are even transformed and obfuscated in the world of commodities is a basis for moral condemnation. But we must be careful here to state the case precisely. It is not primarily the effects that lead to a condemnation of consumption. Rather, it is the act of obfuscating the fact that effects are produced that leads to condemnation. We are on more solid ground condemning obfuscation than the actual effects, since we would have to be able to trace them and decide if they have damaged nature, exploited social relations, and degraded meaning. Determining and assessing these effects are difficult and may lead to often intractable problems. Geography plays an essential role in tracing these connections, but even with the use of place at various scales, it is not always easy to reconstruct the spatial chain of events connecting actions to other actions so that we can describe the consequences of consuming this or that product. Even if a consensus emerged about particular chains of causes and effects, difficulties would still arise over their moral implications due to the fractious state of moral theory. Indeed, many effects may be seen as insignificant or even beneficial. Particular effects, even bad ones, cannot be used to condemn all of mass consumption.

But solid geographical ground does exist for condemning consumption if we shift from its effects on the realms to its obfuscation of effects. Here, the consumer's world undermines our ability to act morally and responsibly because it obscures the consequences of our actions at the very outset. The consumer's world portrays itself as a context without context, a front stage without a backstage. In short, it is geographically disorienting. True, there are many other aspects of our culture that alter our perceptions: we escape reality by going to the movies, reading fiction, and taking drugs. And many of our systems of belief attempt to put reality in a "better" light. But unlike these other aspects of culture, the world of consumption denies the existence of a broader reality. Moreover, consumption is not simply an intellectual system; rather it is a principal means by which we transform the world. Thus we must be able to judge our actions in terms of their effects.

While we do not know precisely what these effects may be, we must nevertheless be aware that they will have effects and attempt to estimate them. And such estimates require that we consider the geographical consequences of our actions, for this is the way we can think of the three realms together and view them from inside and out. (And we can reverse the point of our argument and claim that social theory and morality have both suffered by not considering the geographical context.) The consumer's world is thus immoral, since it creates place out of geographic context and denies us the opportunity of knowing the consequences of our actions. Indeed, it makes it appear as though there is no world supporting consumption and, therefore, no need to understand this world. Mass consumption thus fosters irresponsibility.

Extending Geographical Awareness

One of the seductions of the consumer's world is that it frees us from concern. It is not easy for us to be responsible in this world. Determining the effects of our actions as consumers are so complex that we could become completely paralyzed. This burden, of course, does not exist for consumption alone, but it is consumption that all of us engage in through our everyday activities. How can we know which commodity is the most benign? How can we compare the effects of soil erosion in Colombia due to the cultivation of coffee to the economic impact on Colombia of a slump in the coffee market? We must create institutions to help us understand these effects. Government, scientific, and consumer groups could articulate the consequences of consumption on meaning, nature, and social relations. (Consumer groups do draw attention to these effects by focusing on one or another of the forces: some groups single out environmental degradation, others social injustices, and still others the dilution of meaning.)[50]

Controlling advertising could also help us make moral choices. Advertising could promote an awareness of the consequences of consumption, but only if we were confident that the effects of our actions were understood (see figures 28 and 29). Otherwise, the ads would simply represent opinion, in which case competing advertising campaigns would be launched to present other sides of the issue, and we would be back to the present condition in advertising. In circumstances where everyone agrees about the negative consequences, such as in the consumption of cigarettes, alcohol, and drugs, advertising has already been used to warn consumers. The only element in these ads that differentiates them from others is that they emphasize the generic, rather than the specific, qualities of the commodities. The ads say that all alcohol, all drugs, and so on have this negative effect, not this or that brand. In all other respects, these ads are the same as ads whose aim is to sell a commodity. They retain the tensions of apart/together, we/they, and self/others. This fact—that advertising cannot extricate itself from its own structure—limits the use of advertising for the public good even if there is agreement about a commodity's effects. Advertising simply cannot present information without transforming it into the advertising mode. It might be possible to ban advertising completely, but then we would have to find other means by which meanings can be imparted to vast quantities and varieties of products and communicated to the public.

We might argue that environmental degradation, social exploitation, and immorality are rampant, and we might believe that these are related not only to consumption but to capitalism. Would it not follow that, if we revamp capitalism or destroy it, we will remove the source of the problem? My answer to this is that, though all this may well be true, we have now stepped beyond the common ground and into the conflicts among forces and perspectives; and even if it is not true, I may find capitalism offensive for other reasons. But what I call capitalism and what I think it does and what you call capitalism and what you think it does may differ significantly. Each of our cases may be poorly supported by the facts, so that the argument for change has only weak explanatory force. This does not prevent us from believing in an argument or in acting on these beliefs.

A commitment to change society requires a vision of something else to replace it, and the sharpest of these visions are found in the plans of utopian communities. Utopian experiments have one thing in common: they are attempts at creating places and communities. These places are often used to forge purpose and responsibility. In the more conservative communities, this is accomplished by imposing austerity through limiting production and consumption to the geographical boundaries of this single community. These constraints on production and consumption make it possible for inhabitants of a utopia to see how their actions affect each other and how life's labors promote purpose and responsibility. They remind us that it may be possible to free ourselves from weightlessness by constraining our lives and anchoring them to the necessity of work.

Most utopias, though, are idealistic to the point where there is no bridge between them and the rest of the world. Rather, they simply begin anew by stepping out of society. Most of them offer only local solutions. They are silent about how to create purpose, responsibility, and some measure of austerity in a global society and economy. The most influential exception is the utopian vision of Marx, which is global in outlook and which describes at least some of the steps needed to attain utopia. His view, though, does not point us down the road of austerity but in quite the contrary direction.[51] Far from being an impoverished place, a Marxian utopia supposes an abundance of material wealth resulting from the miracles of technology and transformed social relations. These changes are to eliminate the capitalist division of labor and its alienating work and minimize socially necessary labor time and the need to toil. Because Marx had great faith in human creativity, he believed that we would not be idle even in the midst of plenty. Rather, once we were virtually liberated from necessity, we would lose ourselves in truly meaningful, rewarding work. We would be able to hunt in the morning, fish in the afternoon, and discuss philosophy (be "critical critics") in the evening.[52] Work would be a joy and would express our creativity.

Although work, not consumption (the commodity form of which would disappear), is emphasized in Marx's utopia, his utopia is still based on an abundance of goods. We would consume what we wanted or needed. People who might be lost in their work would be assured that the results of their labor would be accepted by others. Moreover, most of the work would not be driven by necessity; there would be virtually no necessary labor. Rather, most work would be determined by our need for self-fulfillment and self-expression.

However, separating work from necessity runs the risk of altering the character of work to the point that it becomes little more than a diversion or a hobby.[53] When work is divorced from necessity and is nothing more than self-fulfillment, it is in danger of becoming as weightless as consumption. Indeed, there is little to distinguish one from the other. If we have no strong and enduring interests but flit from one hobby to another, our efforts at self-expression are fragmented and we become dilettantes, doing a bit of this and a bit of that. Our efforts at producing become as weightless and shallow as our efforts at consuming. Consumers can, in fact, be called producers when they use consumption to transform themselves and to create contexts, or when they purchase art to build collections, or when they buy model kits to construct antique cars. The Marxian paradise, moreover, does not guarantee that our work would be responsible to nature, meaning, or social relations. Since there is no explicit geography to this paradise, we would not know the consequences of our actions.

Mass consumption rests on abundance, and abundance lessens the weight of necessity in both consuming and producing. These are modern problems, and they are experienced by only a portion of the world. For the multitudes of the poor, necessity and responsibility are borne of grinding poverty. Yet if we are all to have high standards of living, or if this becomes the aspiration of most, the problem of abundance must be confronted. Can a prosperous society introduce austerity? Can work and consumption be given weight by simply defining or dictating tastes and tasks? Can people lose themselves in projects and find them liberating? And even if they can, or if they at least immerse themselves in what Mihaly Csikszentmihalyi calls the "flow" experience of activity, such work might still not be weighty, because it would not be borne out of interest and would have the appearance of a hobby.[54]

Responsibility, and the purpose and weight it entails, can come, however, from an awareness of the geographical conditions associated with consuming and producing. This awareness automatically embraces the connections among the entire range of forces and perspectives. It requires that we imagine ourselves in a place that allows us to see the force of free agency and to see the connections of this place to other places. In this way, we build a picture of the webs of relations that our actions entail, and our awareness of these connections adds restraint. Geographical awareness provides the only common ground. It can be extended but not by theorizing the connections among the realms. Rather, more geographical ground for agreement can develop by examining the relation among forces and perspectives as they are constituted in and by particular geographical places. Recognizing that places are an essential means by which we make sense of the world and through which we act expands the overlap among the realms, geographically integrates theory, and joins it to practice.

· 9 ·

Afterword: Geographical Analysis and the World of Consumption

This book provides a relational framework for geography and its connection to knowledge and action. The framework presents a dynamic picture of how space and place shape—and are shaped by—forces and perspectives, how geography conjoins epistemological and ontological issues, and how it forms a basis for moral judgment. The geographical activity chosen was the world of consumption because it is the most accessible and powerful device by which we daily create places and transform the world. Disclosing how and why consumption is indeed a powerful geographical activity demonstrates the framework's flexibility. The loom, the form the framework takes when applied to consumption, reveals how modern places are woven and unravel and how parts of the relational framework itself are thereby altered.

The framework can be used to explore a wider range of geographical positions and actions than discussed in this book. Focusing on perspectives, we notice that major geographical views can be located along the discursive axis, from the humanistic approach from somewhere, which attempts to see the world from the perspective of experience, to a more distant and decentered chorological view, to the most distant view from the nowhere of spatial analysis. Each view illuminates different aspects of space and place, which then are incorporated into a definition of geography: humanists, with their close-up perspective, see geography as the study of how humans make the earth into a home, chorologists see geography as the study of areal differentiation and integration, and spatial analysts see it as the study of spatial relations. Awareness of the positions from somewhere to nowhere, however, makes it possible to understand how each of these (and still others along different axes) are related perspectives illuminating facets of geographical experience.

Forces, too, can be used to map out geographical positions. Physical and human geography (from any perspective) obviously diverge according to their emphasis on physical and human forces—with the various branches of human geography focusing on particular facets of meaning or social relations, and branches of physical geography dividing along the lines of the major earth forces in the physical world—while geographical ecology attempts to bridge the two. The boundaries and contexts of these divisions change according to our theories about what constitutes the particular forces and their relations.

While geography has always included a range of forces, it tends to emphasize the power of one of these realms over another. Geography has shifted in the last hundred years from an emphasis on the power of nature over social relations (in the various theories of environmentalism) to an emphasis on the effects of social relations and meaning on nature. Most recently, it has stressed the role of agent, particularly in connection with social relations. The framework used here maps out the field of geography, situating positions and actions within the forces and perspectives (and demonstrating how the two are related). It also reveals how a commitment to one or another perspective employs different facets of the framework and illuminates certain qualities of space and place but leaves others in the dark, even implying that they are insignificant or do not exist.

The framework goes beyond the general notion that place integrates everything and anything to the more specific idea that place has a structure based on how it connects meaning, nature, social relations, and agency. Places differ in the mechanisms employed to draw together and alter these forces and their mix. Because we all are involved in the construction and maintenance of places, we are also continuously engaged in debates and struggles over the addition or subtraction of particular forces within them. In changing the mix, we change the character of the place.

For example, wilderness areas emphasize the realm of nature. But debates rage about the meaning we give to that "nature" and the degree to which such places should include or exclude certain types of social relations. If these places become open to oil exploration and drilling, for example, they come to include fewer elements of nature and more elements of particular forms of social relations.

Churches, museums, schools, and universities are supposed to be places that principally emphasize facets of meaning. They are also in natural settings and subject to the forces of nature; they offer people employment, a social relation. Not surprisingly, conflicts arise over the proper mix of these forces for such places. If questions of wages, working conditions, and equal employment opportunity rise to the fore, forces of social relations loom ever larger in these places, displacing and transforming the pursuit of certain facets of meaning.

Even our homes change their mixes of forces. If someone is seriously ill, the home comes too much under the grip of natural forces. If husbands and wives quarrel, if children are abused, the place becomes influenced by certain social relations. And if we teach our children in our homes to explore various ideas and beliefs, our homes then contain certain facets of meaning.

The framework allows us to see how different kinds of place combine forces differently and how places can specialize by depending on one another. Homes need not be completely in the grip of such natural forces as disease, because there are hospitals to take over the care of those who are seriously ill; and homes depend on schools to develop and elaborate certain kinds of meaning and social relations.

The framework thus maps out the structure of places as they combine and recombine different types of forces. Indeed, the framework makes it clear that these forces and the perspectives from which they can be seen are not simply abstractions but are constituted and reconstituted in everyday places and practices. Intentionally or unintentionally maintaining or changing the character of our homes, schools, workplaces, neighborhoods, or states means keeping or altering particular mixes of forces and perspectives. Understanding how places work helps us understand how these forces and perspectives are constantly negotiated, balanced, and transformed.

The relational framework has another merit, or power. It provides a critique of social and moral theories and shows how a geographical common ground can help develop sounder models, linked to everyday practices. To be both empirically competent and morally responsible, we must understand the consequences of our actions. The connections among things in the world must be understood in terms of chains of cause and effect through space and time. Our actions have spatial consequences, yet current theories are too narrow and partial to trace them. The relational framework, however, offers a balanced picture of these consequences, which enables us to trace the impacts of our actions at various scales and perspectives and which helps to build geographically based models and theories.

Equally important, application of the relational framework can help us arrive at an informed consensus about how we ought to act. The framework does not offer specific solutions (in the manner of a handbook) to specific societal and environmental problems, but it does show how informed consensus must be built on an awareness of all the perspectives and all the realms of force—on the entire map of geography. Without the relational framework, our reasons for supporting or resisting a particular action are just as fragmented as our theories.

The fragmentation of theory and, inevitably, of practical response can be seen in current environmental and social attempts at bringing to light the consequences of our actions. Many environmental movements focus on a narrow chain of events within nature, perhaps the particular chemical cycle of a manufacturing process or the effects on a single species of a particular agricultural practice. Even the broadest environmental studies, such as those modeling global warming, still work within the realm of nature and consider human action only in terms of its physical effect on natural systems. Such segmentation is found in social movements that ask consumers to boycott a product because of its connection to certain labor practices or political regimes or because of its implications for particular values and meanings. These protests do not often balance their analyses with a concern for natural processes.

The relational framework provides a critique of these environmental and social movements; it also points us in the direction of responsible policy. Claims for action in an interrelated global system cannot rest on connecting that action to only nature, only meaning, or only social relations. Rather, legitimate claims must consider an action's effects on all three, from the global to the local scale. The foundation for understanding these relations at multiple levels and from different perspectives is what geography provides.

•

Our day-to-day actions affect nature, meaning, and social relations at every step from the local to the global scale. Stating this is equivalent to saying that our actions are geographical and must be understood through geography. Geography explores how we make the earth into a home, and its particular method is the examination of this home-building activity in the context of space and place. Place simultaneously invokes the forces of meaning, nature, social relations, and agency and the perspective from inside (or somewhere) to outside (or nowhere). Creating homes and making them better places require that we understand the relations among forces and perspectives in the context of space and place. Place is both an empirical and a moral concept, and the exercise of the geographical imagination will expand our knowledge of the connections among forces and perspectives and anchor this knowledge to our everyday actions and experiences of being in the world.

Notes

· 1 ·
Introduction: Places of Consumption and the Relational Framework

1. Yi-Fu Tuan, "The Significance of the Artifact," *Geographical Review* 70 (1980): 462–72; and Hannah Arendt, *The Human Condition: A Study of the Central Dilemmas Facing Modern Man* (Chicago: University of Chicago Press, 1958), esp. pp. 120–53, discuss the means by which our physical settings reify and stabilize our transient feelings.

2. See William Butler Yeats, "The Second Coming": "Turning and turning in the widening gyre/The falcon cannot hear the falconer;/Things fall apart; the centre cannot hold;/Mere anarchy is loosed upon the world,/The blood-dimmed tide is loosed, and everywhere/The ceremony of innocence is drowned;/The best lack all conviction, while the worst/Are full of passionate intensity." Also see Karl Marx, "The Manifesto of the Communist Party," in Karl Marx and Friedrich Engels, *Selected Works* (Moscow: Foreign Languages Publishing House, 1955), 1:37:

> The bourgeoisie cannot exist without constantly revolutionising the instruments of production, and thereby the relations of production, and with them the whole relations of society. Conservation of the old modes of production in unaltered form was, on the contrary, the first condition of existence for all earlier industrial classes. Constant revolutionising of production, uninterrupted disturbance of all social conditions, everlasting uncertainty and agitation distinguish the bourgeois epoch from all earlier ones. All fixed, fast-frozen relations, with their train of ancient and venerable prejudices and opinions, are swept away, all new-formed ones become antiquated before they can ossify. All that is solid melts into air, all that is holy is profaned, and man is at last compelled to face with sober senses, his real conditions of life, and his relations with his kind.

3. The power of place to reify leads to paradoxes. One is that, whereas nature is real and preserving it in places such as wilderness and parks may make it more tangible and visible, the very act of preserving nature may also undermine it, because it is then dependent on human actions. This problem leads to the question of the authenticity of place.

4. Tuan, "The Significance of the Artifact," p. 463. Arendt, *The Human Condition,* p. 137, says: "The things of the world have the function of stabilizing human life, and their objectivity lies in the fact that . . . men . . . can retrieve their sameness . . . by being related to the same chair and table. . . . Against the subjectivity of men stands the objectivity of the man-made world rather than the sublime indifference of an untouched nature." Landscapes can stabilize and reify a group's identity and history. See Donald Meinig, ed., *The Interpretation of Or-*

dinary Landscapes: Geographical Essays (New York: Oxford University Press, 1979); John B. Jackson, *American Space: The Centennial Years 1865–1876* (New York: Norton, 1972); John B. Jackson, *The Necessity for Ruins, and Other Topics* (Amherst: University of Massachusetts Press, 1980); John Stilgoe, *Common Landscapes of America 1580–1845* (New Haven: Yale University Press, 1982).

5. James Vance, *This Scene of Man: The Role and Structure of the City in the Geography of Western Civilization* (New York: Harper's College Press, 1977), which was revised as *The Continuing City: Urban Morphology in Western Civilization* (Baltimore: Johns Hopkins University Press, 1990). See also James Vance, *The Merchant's World: The Geography of Wholesaling* (Englewood Cliffs, N.J.: Prentice-Hall, 1970); Michael Chisholm, "The Increasing Separation of Production and Consumption," in B. L. Turner, et al., eds., *The Earth as Transformed by Human Action* (Cambridge: Cambridge University Press, 1990), makes the important point that specialization and trade are very old; he traces their history in terms of how they separate production from consumption.

6. Consider the arguments of Max Weber, *The Protestant Ethic and the Spirit of Capitalism* (New York: Scribner's, 1958); and R. H. Tawney, *Religion and the Rise of Capitalism* (New York: Harcourt, Brace, 1926).

7. Gunnar Olsson, "The Social Space of Silence," *Society and Space* 5 (1987): 249–61, discusses the self-consciousness of postmodernism. See note 12 for other postmodern references.

8. Marshall Berman, *All That Is Solid Melts into Air: The Experience of Modernity* (New York: Simon and Schuster, 1982), pp. 16–17. See his distinctions among *modernity, modernization,* and *modernism. Modernization* refers to social processes and *modernism* to attitudes, values, and beliefs. The term *modernity* includes their dialectical connections.

9. Ibid., p. 17.

10. Ibid., p. 35.

11. Ibid., p. 24.

12. The condition of postmodernity (which could include the social processes of postmodernization and the intellectual processes of postmodernism, to make this condition symmetrical to modernization and modernism) has influenced branches of virtually all the humanities and most of the social sciences. Its origins are in architecture and literary criticism. For advocates and others sympathetic to the position in geography, see Michael Dear, "The Post Modern Challenge: Reconstructing Human Geography," *Transactions, Institute of British Geographers* 13 (1988): 262–74; Michael Dear, "Postmodernism and Planning," *Environment and Planning D: Society and Space* 4 (1986): 367–84; Edward Soja, *Postmodern Geographies: The Reassertion of Space in Critical Social Theory* (London: Verso, 1989). David Ley, "Modernism, Post-modernism, and the Struggle for Place," in John Agnew and James Duncan, eds., *The Power of Place: Bringing Together Geographical and Sociological Imaginations* (London: Unwin-Hyman, 1989), is a particularly insightful analysis of the movement in a historical geographical context. In sympathy with its claims is Dagmar Reichert, "Writing

around Circularity and Self-reference," in Reginald Golledge, Helen Couclelis, and Peter Gould, eds., *A Ground for Common Search* (Goleta, Calif.: Santa Barbara Geographical Press, 1988). Among nongeographical postmodernists who have influenced the field are Jean-François Lyotard, *The Postmodern Condition: A Report on Knowledge* (Minneapolis: University of Minnesota Press, 1984); and Jean Baudrillard, *Simulations* (New York: Semiotext[e], 1983).

13. Bernard Williams, *Ethics and the Limits of Philosophy* (Cambridge: Harvard University Press, 1985), pp. 198–99, argues that critics of the general approach of science have often two motivations:

> On the one hand they say that those who believe that science can tell us how the world really is are superstitiously clutching on to science, in a desperate faith that it is the only solid object left. But equally one may say that comfort is being sought in the opposite direction, and that skepticism against science serves, as it did in the seventeenth century, to warm those whose own claims to knowledge or rational practice look feeble by comparison. The idea that modern science is what absolute knowledge should be like can be disquieting, and it can be a relief if one represents science as merely another set of human rituals.

Quoted, and discussed in a geographical context, in J. Nicholas Entrikin, *The Betweenness of Place: Toward a Geography of Modernity* (Baltimore: Johns Hopkins University Press, 1990).

14. See Thomas Nagel, *The View from Nowhere* (New York: Oxford University Press, 1986), esp. pp. 90–99, for a discussion of the interrelationships among realism, idealism, and skepticism. The only form of realism that geographers now discuss is transcendental realism, drawn from Roy Bhaskar, *A Realist Theory of Science* (Leeds, Kent: Alma, 1975). Also see Andrew Sayer, *Method in Social Science: A Realist Approach* (London: Hutchinson, 1984). See also the discussion about realism and verification in Robert Sack, "Realism and Realistic Geography," *Transactions, Institute of British Geographers* 7 (1982): 504–9, and the excellent review of geography's handling of this debate in Entrikin, *The Betweenness of Place.*

15. It is possible to be in sympathy with postmodernism and still believe that local or contextual truths could emerge and become the basis for local communities. These "plural styles" are the basis of postmodern architecture; see Ley, "Modernism, Post-modernism, and the Struggle for Place," p. 53. But the postmodernist conception of this plurality of perspectives is not a stable position, because the "local community" is largely ephemeral in modern society; multiple perspectives become the chaotic, quixotic relativity of postmodernity.

16. David Harvey, *The Condition of Postmodernity: An Enquiry into the Origins of Cultural Change* (Oxford: Basil Blackwell, 1989), p. 52.

17. Berman, *All That Is Solid Melts into Air,* p. 15.

18. Harvey, *The Condition of Postmodernity,* p. 44.

19. Ibid., p. 51.

20. Ibid., p. 98.

21. Harvey, *The Condition of Postmodernity* is the most insightful overview of the movement to date, yet its reliance on Marxist interpretation of modernity and on the effects of flexible accumulation (see note 22) on postmodernity situates the origins of this movement in a vulnerable and narrow context. See also David Harvey, "Flexible Accumulation through Urbanization," *Antipode* 19 (1987): 260–86.

22. Frederic Jameson, "Postmodernism, or the Cultural Logic of Late Capitalism," *New Left Review* 146 (1984): 53–92; Harvey, *The Condition of Postmodernity*. Jameson and Harvey attempt to explain the phenomenon as a result of capitalism, Harvey seeing it as a development in the capitalist mode of production called flexible accumulation. The concept draws on a narrower and more specific set of relations that go under the terms *flexible production* and *flexible specializations*. But how do these work? On this matter, there is much debate and uncertainty. For example, Michael Piore and Charles Sabel, *The Second Industrial Divide: Possibilities for Prosperity* (New York: Basic Books, 1984), argue that flexible specialization increases the need for workers' skills. Harley Shaiken, Stephen Herzenberg, and Sarah Kuhn, "The Work Process under More Flexible Production," *Industrial Relations* 24 (1986): 167–83, argue that it does not.

Scott and Cooke claim that flexibility is found "in such novel phenomena as (a) CAD-CAM production systems, and robotized work stations, (b) just-in-time processing of inputs . . . and (c) in many restructured mass-production sectors, flexibly specialized assembly lines and neo-fordist experiments with work teams, quality circles, job-enrichment programs, and the like"; A. J. Scott and P. Cooke, "The New Geography and Sociology of Production," *Environment and Planning D: Society and Space* 6 (1988): 241–44. They claim this will result in the reduction in time to retool and thus lead to a shift from assembly line mass production to batch production and, hence, to a greater variety of products and rapidity of change in products. This means greater options for consumers and, thus, an even greater investment in advertising.

Flexibility in production could also allow a firm greater options in the realm of accumulation. Companies could shift rapidly among economic sectors and, eventually, even among circuits of capital, so that a company could almost instantly shift its capital among its interests in production, the built environment, and finance. This potential could lead to a blurring of distinctions among the circuits, allowing the very largest multinational companies to embrace all three. (The U.S. attempt in the nineteenth century to separate the banking world from the world of production is already weakening.)

Accompanying flexible production and accumulation is the possibility of increased geographical flexibility. This could occur because these new means of production and accumulation allow capital to become more mobile and even footloose. Capital would become more multinational (see, for example, the *New York Times* 21 May 1989) and freer to move among different geographical mixes of factors of production and market accessibility. The increase in geographical mobility of capital could create a geographical mismatch between capital and national territories. See Robert Sack, *Human Territoriality: Its Theory and His-*

tory (Cambridge: Cambridge University Press, 1986); and Michael Storper and Allan Scott, "The Geographical Foundations and Social Regulation of Flexible Production Complexes," in J. Wolch and M. Dear, eds., *Territory and Social Production* (London: Allen and Unwin, 1988).

This would limit a state's capacity to control business and to provide a welfare function and would place a greater burden on political entities to compete in attracting and retaining capital. This means that the local and national states would increase their involvement in the creation of attractive infrastructures, investment incentives, subsidies, and partnerships. According to Harvey, *The Condition of Postmodernity,* p. 156:

> Flexible accumulation has been accompanied on the consumption side . . . by a much greater attention to quick-changing fashions and the mobilization of all the artifices of need inducement and cultural transformation that this implies. [This has led to heightened] ferment, instability, and fleeting qualities of a post-modernist aesthetic that celebrates difference, ephemerality, spectacle, fashion, and the commodification of cultural forms.

This is also accompanied by a much more fragmented and unevenly developed space economy.

N. Albertsen, "Postmodernism, Post-Fordism, and Critical Social Theory," *Environment and Planning D: Society and Space* 6 (1988): 339–65, extends this even further by claiming that "everyday life is becoming thoroughly aestheticized . . . as use values increasingly give way to signs, symbols, [and] images of pleasure in consumption, and as the avant-garde assumed the role of the R&D of capitalist commodity aesthetics and production."

Whether flexible production and accumulation do in fact have clearly discernible geographical consequences, whether these are actually novel ones, as implied by the argument that flexible accumulation creates postmodern landscapes, and whether even flexible accumulation follows from flexible production remains to be seen.

Still, it seems possible that flexible production and accumulation could lead to an acceleration in types of commodities, rates of consumption, and an acceleration in combined state and industry investment in the creation of places as contexts for consumption. Thus the scale and rapid turnover in such investments make places ever more fluid. This alone could heighten the tensions and contradictions that these places possess. But the critical point is that places possess them not by virtue of flexible accumulation, but by virtue of the fact that they contain commodities and are themselves commodities. Commodities are, of course, part of capitalism, but capitalism can be as much a matter of meaning as of particular social relations.

For a more skeptical view about the geographical effects of flexible accumulation, see Meric Gertler, "The Limits to Flexibility: Comments on the Post-Fordist Vision of Production and Its Geography," University of Toronto, Department of Geography. See also Andrew Sayer, "Post-Fordism in Question," *International*

Journal of Urban and Regional Research 13 (1989): 666–95. (I am indebted to Chris Thompson for several of these references.)

23. These activities are described as industries, and the brochures they publish emphasize the fact that they are to be mass consumed. A tourist guide for Florida is called "Consumer's Guide to Florida."

24. The term *personal place* here is different from the personal distances and space discussed by writers such as Edward Hall and David Stea and reviewed by J. Douglas Porteous, *Environment and Behavior: Planning and Everyday Urban Life* (Reading, Mass.: Addison-Wesley, 1977), pp. 26–59.

25. Although space and place have many meanings, they are dialectically related. Yi-Fu Tuan, *Space and Place: The Perspective of Experience* (Minneapolis: University of Minnesota Press, 1977), provides the definitive discussion of the relation between space and place from the perspective of experience. At this level, place can be thought of as a locus of attention, whereas space can be associated more with movement. Place precedes space in prominence. Space, in fact, can be thought of as a series of places through which one moves.

26. For a discussion of portions of the geographical history of space, see Sack, *Human Territoriality;* Harvey, *The Condition of Postmodernity,* esp. pp. 201–308; and Derek Gregory, "Space and Time in Social Life," Wallace W. Atwood Lecture Series, Clark University, 1985.

27. For two complementary analyses of these perspectives, see Entrikin, *The Betweenness of Place;* and Robert Sack, *Conceptions of Space in Social Thought: A Geographic Perspective* (Minneapolis: University of Minnesota Press, 1980).

28. Nagel, *The View from Nowhere.* Nagel does not link the concepts of public and private to these distinctions, because he wishes to draw attention to the possibility that those things that can be thought of or felt subjectively by one person can also be apprehended (and thus thought of and felt) by others—perhaps through phenomenological investigations—and thus these thoughts and feelings are not really private. I agree, but I link the private with the personal and subjective and the public with the abstract and objective, because that is the way contemporary society views these feelings. Even if introspection allows us to know the feelings of others, our culture does not treat this knowledge with the same degree of confidence as it does the more objective knowledge. This is why we see the subjective as more closely associated with the private, and the objective with the public. See chapter 3 for greater elaboration of subjective/objective.

29. Jean Piaget and Barbel Inhelder, *The Child's Conception of Space* (New York: Norton, 1967).

30. Giddens uses the term *structure* in the context of social relations, but it can also be used to describe the rules of transformation of psychological forces or nature. See Anthony Giddens, *The Constitution of Society: Outline of the Theory of Structuration* (Berkeley and Los Angeles: University of California Press, 1984), esp. p. 17.

31. For a sympathetic overview of the problem of structuration, see Ira Cohen, *Structuration Theory: Anthony Giddens and the Constitution of Social Life* (New York: St. Martin's, 1989), esp. pp. 11, 23–26, and 219–28. I focus on

choice as a demonstration of will, rather than as the creation and execution of projects, because this is the way the problem is usually posed by social science.

32. The phrase is from Yi-Fu Tuan's "Space and Place: A Humanistic Perspective," in Stephen Gale and Gunnar Olsson, eds., *Philosophy in Geography* (Dordrecht, The Netherlands: D. Reidel, 1979), pp. 387–427.

33. See Sack, *Conceptions of Space,* for a more detailed discussion of perspectives and modes of thought.

34. Somewhere and nowhere are linked to subjectivity and objectivity. See Nagel, *The View from Nowhere;* and Sack, *Conceptions of Space.* Definitions of geography correspond to lenses on the discursive/scientific axis. The close-up, humanistic view sees geography as the transformation of the earth into a home for human beings. The distant, nomothetic, positivistic view sees it as the study of spatial relations. In between are other lenses, such as chorology, which views geography as the study of areal integration and differentiation.

35. Immanuel Kant, *Critique of Pure Reason* (London: Macmillan, 1950); and P. F. Strawson, *Individuals: An Essay in Descriptive Metaphysics* (New York: Anchor/Doubleday, 1963).

36. Soja, *Postmodern Geographies,* esp. pp. 43–76; and David Harvey, *The Limits to Capital* (Chicago: University of Chicago Press, 1982), chap. 13.

37. In the theory of territoriality, influence and power are neutral in that they are necessary components to all action. Even conversation is an attempt to affect or influence others and may require a territorial structure such as a room. See Sack, *Human Territoriality.*

38. Harvey, *The Limits to Capital;* and Michael Mann, *The Sources of Social Power* (Cambridge: Cambridge University Press, 1986).

39. Territoriality can intensify and differentiate areal differences and create new ones. In Robert Sack, "The Consumer's World: Place as Context," *Annals of the Association of American Geographers* 78 (1988): 642–64, territorial segmentation and integration, instead of areal differentiation and integration, are mistaken as the threads drawn from nature.

· 2 ·
Perspectives from Somewhere to Nowhere

1. In Robert Sack, *Conceptions of Space in Social Thought: A Geographic Perspective* (Minneapolis: University of Minnesota Press, 1980), science, social science, art, myth and magic, the child's perspective, and the practical perspective are explored historically.

2. The terms *personal space* and *personal place* are used elsewhere to refer to a fixed area surrounding the individual; see J. Douglas Porteous, *Environment and Behavior: Planning and Everyday Urban Life* (Reading, Mass.: Addison-Wesley, 1977), esp. pp. 31–59. This is a view from outside, or nowhere, rather than inside, or somewhere.

3. Humanistic geography has many meanings and draws on several philo-

sophical traditions, including phenomenology and existentialism. For a review of this area of geography, see David Ley and Marwyn Samuels, eds., *Humanistic Geography: Prospects and Problems* (Chicago: Maaroufa, 1978); and John Pickles, *Phenomenology, Science, and Geography* (Cambridge: Cambridge University Press, 1984). The philosopher Hao Wang, *Beyond Analytic Philosophy: Doing Justice to What We Know* (Cambridge: MIT Press, 1986), argues for a descriptive introspective method that is not phenomenology but rather "phenomenography." He claims that

> phenomenography rejects the . . . emphasis on subjectivity to the exclusion of objectivity and even intersubjectivity, and it rejects the search for an "a priori formulation once and for all." Rather it focuses on both what we know and how we feel, and the analogy with geography is intentional —suggesting a 'quest for a structural comprehensiveness' rather than . . . attention to limited aspects or domains required of specialized disciplines. (p. 37)

This idea of place employs phenomenography and simple introspection. It asks, What are the necessary and essential geographical conditions accompanying being in place? One might contend that this approach is part realist in that it, too, asks what humans must be like for certain actions to occur.

4. This is similar to T. S. Eliot's statement in "Little Gidding" that only after you have returned do you know the place you have left.

5. These dependencies allow our personal sense of place to merge with the public sense of place and space. The details of how this occurs and how personal place is linked to social routines, practices, and structures, can be based on Anthony Giddens' work on "distanciation" and "co-presence." See his *The Constitution of Society: Outline of the Theory of Structuration* (Berkeley and Los Angeles: University of California Press, 1984); Theodore Schatzki's work on spatiality ("Social Spatiality," University of Kentucky, Department of Philosophy); and the time geographical research of Torsten Hagerstrand, "Dioram, Pand and Project," *Tijdschrift Voor Economishe en Social Geographific* 73 (1982): 323–39. For a discussion of the role of spatial routine in instantiating authority, see Michel de Certeau, *The Practice of Everyday Life* (Berkeley and Los Angeles: University of California Press, 1984); and Pierre Bourdieu, *Outline of a Theory of Practice* (Cambridge: Cambridge University Press, 1977).

6. Resistance to routine often requires disclosing the taken-for-granted spatial structures, which are supported by territoriality. This can occur by using precisely those advantages that territoriality provides and that are predicted by the theory; see Robert Sack, *Human Territoriality: Its Theory and History* (Cambridge: Cambridge University Press, 1986), but this time in reverse. Graffiti and carnivals, for example, both disclose and resist the boundaries that are taken for granted. See Peter Jackson, "Street Life: The Politics of Carnival," *Society and Space* 6 (1988): 213–27; and Peter Jackson, "Social Geography: Social Struggles and Spatial Strategies," *Progress in Human Geography* 12 (1988): 263–69. See

also Stuart Hall and Tony Jefferson, eds., *Resistance through Rituals* (London: Hutchinson, 1976).

7. For a discussion of the layers of awareness regarding place, see Yi-Fu Tuan, "A Sense of Place," in Gretchen Holstein Schoff and Yi-Fu Tuan, *Two Essays on a Sense of Place* (Madison: Wisconsin Humanities Committee, 1989).

8. Yi-Fu Tuan, *Space and Place: The Perspective of Experience* (Minneapolis: University of Minnesota Press, 1977), p. 35.

9. See Thomas Nagel, *The View from Nowhere* (New York: Oxford University Press, 1986), p. 4, for a more detailed description. I owe a great debt to Nagel for his analysis of this process.

10. Ibid., p. 108.

11. Sack, *Conceptions of Space.*

12. For a discussion of the importance of the generic versus the specific in geography, see Richard Hartshorne, *The Nature of Geography: A Critical Survey of Current Thought in the Light of the Past* (Lancaster, Pa.: Association of American Geographers, 1939; reprinted 1961); J. Nicholas Entrikin, *The Betweenness of Place: Toward a Geography of Modernity* (Baltimore: Johns Hopkins University Press, 1990); and Robert Sack, "The Nature in Light of the Present," in J. Nicholas Entrikin and Stanley Brunn, eds., *Reflections of Richard Hartshorne's* The Nature of Geography, Occasional Paper (Washington, D.C.: Association of American Geographers, 1989).

13. *Nowhere* and *objectivity,* and *somewhere* and *subjectivity,* are relative terms; no one has pointed this out more eloquently than Nagel, *The View from Nowhere.* If either were absolute, we would encounter what Leszek Kolakowski calls a metaphysical horror. The two poles of this axis of horrors are the absolute and the self, or *cogito.* "Both are supposed to be bastions that shelter the meaning of the notion of existence. The former, once we try to reduce it to its perfect form, uncontaminated by contact with any less sublime reality, turns out to pass away into nothingness. The latter, on closer inspection, seems to suffer the same fate." Leszek Kolakowski, *Metaphysical Horror* (Oxford: Basil Blackwell, 1988), p. 56. The same can be said about absolute free will and absolute determinism.

14. W. Schivelbusch, "Railroad Space and Railroad Time," *New German Critique* 14 (1978): 31–40, is an excellent exposition of how our sense of space has been transformed by transportation systems. The railroad was the first mode to insulate the traveler from the sights, sounds, and smells of the environment, and even from the feel of the ground. It made the experience seem like a duration of time. See also Donald Lowe, *The History of Bourgeois Perception* (Chicago: University of Chicago Press, 1982), and Stephen Kern, *The Culture of Time and Space 1880–1918* (Cambridge: Harvard University Press, 1983).

15. Martin Buber, "Distance and Relation," *Psychiatry* 20 (1957): 97–104; quotation on p. 99.

16. Nagel, *The View from Nowhere,* pp. 3–5.

17. Ibid., p. 68. See also Kolakowski, *Metaphysical Horror.*

18. Nagel, *The View from Nowhere,* p. 63. The work of Jurgen Habermas, *The Theory of Communicative Action,* vols. 1 and 2, trans. Thomas McCarthy

(Boston: Beacon, 1987), provides a more detailed description of how such an objective position is possible through the give and take of enlightened conversation. Glenn Tinder, *Community: Reflections on a Tragic Ideal* (Baton Rouge: Louisiana State University Press, 1980), sees communication and conversation as the ideal of community.

19. William Stern, *Psychology of Early Childhood* (New York: H. Holt, 1924), p. 46, as quoted in Heinz Werner, *Comparative Psychology of Mental Development* (New York: International Universities Press, 1973), pp. 64–65.

20. This definition of symbol is discussed in Sack, *Conceptions of Space,* and is drawn from Susanne Langer, *Philosophy in a New Key: A Study in the Symbolism of Reason, Rite, and Art* (Cambridge: Harvard University Press, 1942); and Ernst Cassirer, *The Philosophy of Symbolic Forms,* 3 vols. (New Haven: Yale University Press, 1968).

21. There are factors in symbol formation that operate in the other direction, making the separation only partial and tentative. The mechanisms countering the separation, fusing the subjective and the objective, are seen in the earliest stages, where symbols often seem to possess both objective and subjective attributes of the things they represent. See Sack, *Conceptions of Space.*

22. Jean Piaget, *The Construction of Reality in the Child* (New York: Basic Books, 1954), p. 15; for the way manipulation is central to the conceptual development of the child, see pp. 19, 20.

23. Jean Piaget and Barbel Inhelder, *The Child's Conception of Space* (New York: Norton, 1967), p. 454.

24. The development of a sense of Euclidean space is paradoxical. On the one hand, it is attained through our manipulations of objects in space. Yet, once it is internalized, it can be conceptually divorced from the very objects and relations that gave it form. Moreover, it does not recapitulate the order of the stages (topology, projective, Euclidean geometry) that are part of its development. Indeed, as Piaget has noted, formal mathematics actually reverses the process in instruction, by teaching Euclidean geometry first, then projective, and finally, topological.

25. See Hartshorne, *The Nature of Geography,* for a discussion of the concepts of areal differentiation and integration.

26. For examples of geographical location theories, see Peter Haggett, Andrew Cliff, and Allan Frey, *Locational Analysis in Human Geography,* 2d ed. (London: Edward Arnold, 1977). Those location theories that address something akin to territoriality are the administrative principles in central place theory.

27. Yi-Fu Tuan, *Segmented Worlds and Self: Group Life and Individual Consciousness* (Minneapolis: University of Minnesota Press, 1982). The theme of the book is "spatial segmentation in relation to developing consciousness and to the idea of self," p. 3.

28. Ibid., p. 9.

29. Ibid., p. 141.

30. Dominique Zahan, *The Religion, Spirituality, and Thought of Traditional*

Africa (Chicago: University of Chicago Press, 1979), p. 8, as quoted in Tuan, *Segmented Worlds and Self*, p. 141.

31. Tuan, *Segmented Worlds and Self*, p. 141.

32. Ibid., p. 155; Tuan quotes Hans Zonas.

33. Ibid., p. 56.

34. Phillipe Aries, *Centuries of Childhood: A Social History of Family Life* (New York: Vintage, 1965), p. 395, quoted in Tuan, *Segmented Worlds and Self*, p. 74.

35. Sack, *Human Territoriality*, p. 89.

36. Tuan, *Segmented Worlds and Self*, p. 78.

37. Ibid., p. 82. Not only were houses divided into specialized rooms but the furniture became specialized and varied; chairs were padded and moved away from the walls to stand in their own places. "The interior furniture of the houses appeared together with the interior furniture of the mind," p. 83.

38. Ibid., p. 133.

39. Ibid., p. 139.

40. Sack, *Conceptions of Space*, and Sack, *Human Territoriality;* David Harvey, *Consciousness and the Urban Experience: Studies in the History and Theory of Capitalist Urbanization* (Baltimore: Johns Hopkins University Press, 1985), pp. 1–35; David Harvey, *The Condition of Postmodernity* (Oxford: Basil Blackwell, 1989), pp. 201–328; and Derek Gregory, "Space and Time in Social Life," Wallace W. Atwood Lecture Series 1 (Worcester, Mass.: Clark University Press, 1985). This discussion draws most on my previous work on the development of abstract geographical space developed in *Human Territoriality*.

41. The point is most controversial in extreme Marxist positions, where everything is determined by the mode of production. But some Marxist theorists (especially cultural Marxists) wrestle with the idea of according meaning almost as much weight as social relations. See the "new cultural geography" of Peter Jackson and Susan Smith, *Exploring Social Geography* (London: Allen and Unwin, 1984); Dennis Cosgrove and Steven Daniels, eds., *The Iconography of Landscape: Essays on the Symbolic Representation, Design and Use of Past Environments* (Cambridge: Cambridge University Press, 1988); and Dennis Cosgrove and Peter Jackson, "New Directions in Cultural Geography," *Area* 19 (1987): 95–101. Perhaps the most insightful analysis of the importance of culture to an economy is Marshall Sahlins, *Culture and Practical Reason* (Chicago: University of Chicago Press, 1976), who argues that the economic realm is driven by symbolic meaning and rules.

42. David Harvey, *Consciousness and the Urban Experience: Studies in the History and Theory of Capitalist Urbanization* (Baltimore: Johns Hopkins University Press, 1985), p. 4.

43. Money is only one of many quantitative forms that lead to abstraction. Harvey, (ibid.) emphasizes money; so does Alfred Sohn-Rethel, *Intellectual and Manual Labor* (London: Macmillan, 1978).

44. The differences between modern and premodern conceptions of time are

discussed by Lewis Mumford, *Technics and Civilization* (New York: Harcourt, Brace, Jovanovich, 1934), esp. pp. 9–28.

45. Joseph Needham, with Wang Ling, *Science and Civilization in China* (Cambridge: Cambridge University Press, 1959), 2:279–301. See also Paul Wheatley, *The Pivot of the Four Quarters: A Preliminary Enquiry into the Origins and Character of the Ancient Chinese City* (Chicago: Aldine, 1971), who discusses the relation between symbolic geography and political integration.

46. These points about Greek conception of space are made by William Ivins, *Art and Geometry: A Study in Space Intuitions* (New York: Dover, 1964); Max Jammer, *Concepts of Space* (New York: Harper and Brothers, 1969), who states that space was conceived by classical Greek philosophy and science at first as something "inhomogeneous . . . and later as something anisotropic . . . [and] these doctrines account for the failure of mathematics, especially geometry, to deal with space as a subject of scientific inquiry," p. 23; O. Neugebauer, *The Exact Sciences in Antiquity,* 2d ed. (Providence: Brown University Press, 1957), p. 225.

47. Samuel Edgerton, "From Mental Matrix to Christian Empire," in David Woodward, ed., *Art and Cartography* (Chicago: University of Chicago Press, 1987).

48. Ibid.

49. For a general discussion of the changing geography of Western cities, see James Vance, *This Scene of Man: The Role and Structure of the City in the Geography of Western Civilization* (New York: Harper's College Press, 1977), which was revised as *The Continuing City: Urban Morphology in Western Civilization* (Baltimore: Johns Hopkins University Press, 1990); and for the transition to the modern city, see David Ward, *Cities and Immigrants: A Geography of Change in Nineteenth-Century America* (New York: Oxford University Press, 1971).

50. Robin Evans, "Figures, Doors, and Passages," *Architectural Design* 48 (1978): 267–78, discusses the interior differentiation of the house.

51. See ibid.; and Tuan, *Segmented Worlds and Self.*

52. David Harvey, *The Condition of Postmodernity: An Enquiry into the Origins of Cultural Change* (Oxford: Basil Blackwell), p. 254.

· 3 ·
The Problem of Agency

1. Susanne Langer, *Mind: An Essay on Human Feeling* (Baltimore: Johns Hopkins Press, 1967), 1:23–24. Jerome Kagan, *The Nature of the Child* (New York: Basic Books, 1984), calls autogenic sensations "internal tone" and the sense of impact "feeling states."

2. See Robert Sack, *Conceptions of Space in Social Thought: A Geographic Perspective* (Minneapolis: University of Minnesota Press, 1980). I also recognize

that this distinction may not stop the circularity of defining that which is objective by the subject matter of the most objective field—i.e., science.

3. The example is taken from Thomas Nagel, *What Does It All Mean? A Very Short Introduction to Philosophy* (New York: Oxford University Press, 1987), p. 50. The idea of freedom can be approached in other ways. For example, think of freedom as the ability to do what you feel you must do (or in Kantian terms, to submit to the laws one gives oneself). This makes freedom the fulfillment of necessity.

4. Reasons and causes are similar. See Donald Davidson, "Actions, Reasons, and Causes," in his *Essays on Actions and Events* (Oxford: Oxford University Press, 1980), pp. 3–19. The two are equated by Anthony Giddens, *New Rules of Sociological Method: A Positive Critique of Interpretative Sociologies* (London: Hutchinson, 1976), pp. 84–85; and Anthony Giddens, *The Constitution of Society: Outline of the Theory of Structuration* (Berkeley and Los Angeles: University of California Press, 1984), p. 345. For the opposite view, see A. R. Louch, *Explanation and Human Action* (Berkeley and Los Angeles: University of California Press, 1966), esp. chap. 6.

5. Nagel, *What Does It All Mean?* esp. pp. 50–52.

6. Ibid., esp. p. 55.

7. Max Rheinstein, ed., *Max Weber on Law in Economy and Society* (New York: Clarion/Simon and Schuster, 1967); Talcott Parsons, ed., *Max Weber: The Theory of Social and Economic Organization* (New York: Free Press, 1947); and Roberto Michels, *Political Parties* (Glencoe, Ill.: Free Press, 1949).

8. Fred Lukermann, "The 'Calcul Des Probabilités and the Ecole Française De Géographie." *Canadian Geographer* 9 (1965): 128–37, is the first American geographer to write of the distinction between *caused* and *determined.*

9. Ibid.

10. Economists would use consumption possibility curves and indifference curves to calculate these preferences.

11. In Giddens, *The Constitution of Society,* the terms *free agency, free will,* and *choice* do not appear as topics. He skirts the issue by decentering the agent.

12. Thomas Nagel, *The View from Nowhere* (New York: Oxford University Press, 1986), p. 110. As Nagel points out here and also in *What Does It All Mean?* pp. 56–57, either possibility raises the alarming situation of our not being responsible for our actions.

· 4 ·
Forces from the Realms of Meaning, Nature, and Social Relations

1. The assumption that virtually everything is eventually connected to everything else is a strongly held belief in the modern world, especially in ecology and environmentalism, in spite of the fact that modern society has fragmented and

compartmentalized thought and action. In tightly woven societies that are closer to nature, the assumption of integration may not be as important.

2. The classic statement of reduction is found in Herbert Feigl and May Brodbeck, eds., *Readings in the Philosophy of Science* (New York: Appleton-Century-Crofts, 1953); May Brodbeck, ed., *Readings in the Philosophy of the Social Sciences* (New York: Macmillan, 1968); and Ernst Nagel, *The Structure of Science: Problems in the Logic of Scientific Explanation* (New York: Harcourt, Brace, World, 1961). Two forms of reduction can be identified: definitional reduction and explanatory reduction. *Definitional* means that the terms and concepts of one science can be reduced to the terms and concepts of another; *explanatory* means that the laws and theories of one science can be replaced by the laws and theories of another. Definitional reduction does not imply that explanatory reduction must occur, although explanatory reduction presupposes definitional reduction. The existence of emergent properties and entelechies make definitional reduction impossible.

3. It is important to remember that when Giddens says that structures help constrain and enable, the enabling is still not equivalent to free agency, because it still would be caused. Now some may argue that structure and agency make freedom unnecessary because we cannot think of being free without structures enabling us. The question of interpreting the meaning of choice—of deciding what it means to say we could have done otherwise—is simply not addressed by enabling. See Anthony Giddens, *New Rules of Sociological Method: A Positive Critique of Interpretative Sociologies* (London: Hutchinson, 1976); and Anthony Giddens, *The Constitution of Society: Outline of the Theory of Structuration* (Berkeley and Los Angeles: University of California Press, 1984).

4. See Lynn White, "The Historical Roots of Our Ecological Crisis," *Science* 155 (1967): 1203–7, reprinted in Francis Schaefer, ed., *Pollution and the Death of Man: The Christian View of Ecology* (Wheaton, Ill.: Tyndall House, 1970). The following is taken from White, "The Historical Roots," pp. 106–15.

5. See the discussion of reasons as causes in chapter 3, note 4. See also Robert Brown, *Explanation in Social Science* (Chicago: Aldine, 1963), pp. 99–108.

6. Yi-Fu Tuan, "Discrepancies between Environmental Attitude and Behavior: Examples from Europe and China," *Canadian Geographer* 12 (1968): 176–91.

7. Claude Lévi-Strauss, *Structural Anthropology* (New York: Oxford University Press, 1963), provides an overview of his method and philosophy. See Edmund Leach, *Culture and Communication: The Logic by Which Symbols Are Connected* (Cambridge: Cambridge University Press, 1976), for a discussion of Lévi-Strauss's methods.

8. G. S. Kirk, *Myth: Its Meaning and Functions in Ancient and Other Cultures* (Cambridge: Cambridge University Press, 1970), p. 44.

9. Leach, *Culture and Communication,* p. 5.

10. Yi-Fu Tuan, *Man and Nature* (Washington, D.C.: Association of American Geographers, 1971); see also Linda Graber, *Wilderness as Sacred Space* (Washington, D.C.: Association of American Geographers, 1976).

11. Tuan, *Man and Nature.*

12. Others have attempted to make structuralism dynamic. See Marshall Sahlins, *Historical Metaphors and Mythic Realities* (Ann Arbor: University of Michigan Press, 1981); and Marshall Sahlins, *Islands of History* (Chicago: University of Chicago Press, 1985).

13. Sigmund Freud, *Civilization and Its Discontents* (New York: Norton, 1961).

14. Calvin Hall and Gardner Lindzey, *Theories of Personality* (New York: John Wiley, 1978), pp. 36–37.

15. See David Lowenthal, "Awareness, Human Impacts: Changing Attitudes and Emphasis," in B. L. Turner et al., eds., *The Earth as Transformed by Human Action* (Cambridge: Cambridge University Press, 1990), pp. 121–36.

16. Edward Wilson, *Sociobiology: The New Synthesis* (Cambridge: Harvard University Press, 1975). See also Arthur Caplan, ed., *The Sociobiology Debate: Readings on Ethical and Scientific Issues* (New York: Harper and Row, 1978).

17. Clarence Glacken, *Traces on the Rhodian Shore: Nature and Culture in Western Thought from Ancient Times to the End of the Eighteenth Century* (Berkeley and Los Angeles: University of California Press, 1967).

18. Larry Grossman, "Man-Environment Relations in Anthropology and Geography," *Annals of the Association of American Geographers* 67 (1977): 126–44.

19. Roy Ellen, *Environment, Subsistence, and System: The Ecology of Small-Scale Social Transformations* (Cambridge: Cambridge University Press, 1982).

20. Glacken, *Traces on the Rhodian Shore,* pp. 42–44, 403–4.

21. J. E. Lovelock, *Gaia: A New Look at Life on Earth* (Oxford: Oxford University Press, 1979).

22. Ellen, *Environment, Subsistence, and System,* p. 76. I owe a great debt to Oliver Coombs's discussions on ecology in "Concept of Cultural Adaptation in Cultural Ecology: A Critical Appraisal of Functional and Strategic Interpretation," University of Wisconsin—Madison, Department of Geography, 1978.

23. See J. Bennett, *The Ecological Transition: Cultural Anthropology and Human Adaptation* (New York: Pergamon, 1976); and T. Earle, "Comment," *Current Anthropology* 25 (1984): 406–7. Interpretations of ecological adaptation are often baffling; two important ones are functional adaptation, usually part of ecosystemism, and strategic adaptation, a part of adaptive dynamics.

24. Ellen, *Environment, Subsistence, and System.*

25. Ibid., p. 122. Allan Schnaiberg, *The Environment: from Surplus to Scarcity* (New York: Oxford University Press, 1980), pp. 18–19, discusses human and other biological systems in terms of energy. One important way human systems diverge from other biological systems is in the

> creation and disposition of surplus energy. . . . [If] the ecosystem changes over time from [a] simpler, faster-growing one to a more complex, slower-growing entity, almost the reverse is true of human economies. . . . Whereas the ecosystem reaches a steady-state by permitting the growth of just enough species and populations to offset the surplus, societies tend to use the surplus

to accumulate still more economic surplus in future periods. . . . Thus societies operate to multiply their surpluses, particularly industrial capitalist societies. In contrast, ecosystems tend to mature by stabilizing numbers of consumers and levels of consumption.

26. I have to use this awkward term because the simpler term *social theory* may not suggest that we are also including theories of the mind; furthermore, it has been appropriated by Marxist and critical theorists.

27. "Multiply its surpluses" is from Schnaiberg, *The Environment.*

28. William Leiss, *The Limits to Satisfaction: An Essay on the Problem of Needs and Commodities* (Toronto: University of Toronto Press, 1976).

29. Capitalism might not need to grow in order to survive, although this extremely important issue is rarely discussed explicitly by contemporary economic theory. The issue has been raised by Malthus and Adam Smith, but modern economics is largely silent on the subject.

30. On the sociology of knowledge, see Karl Mannheim, *Ideology and Utopia: An Introduction to the Sociology of Knowledge* (London: Routledge and Kegan Paul, 1936). On the social construction of reality, see Peter Berger and Thomas Luckmann, *The Social Construction of Reality: A Treatise in the Sociology of Knowledge* (New York: Doubleday, 1966). Jurgen Habermas, *The Theory of Communicative Action,* vols. 1 and 2 (Cambridge: Polity, 1984, 1987), also attempts to bind meaning to social relations. See Derek Gregory, "The Crisis of Modernity? Human Geography and Critical Social Theory," in Nigel Thrift and Richard Peet, eds., *New Models in Geography: The Political-Economy Perspective,* vol. 2 (London: Unwin-Hyman, 1989), pp. 348–85 , for a discussion of Habermas and geography.

31. Alfred Schmidt, *The Concepts of Nature in Marx* (London: New Left Books, 1971), p. 30.

32. Barry Hindess and Paul Hirst, *Pre-Capitalist Modes of Production* (London: Routledge and Kegan Paul, 1975).

33. Karl Wittfogel, *Oriental Despotism: A Comparative Study of Total Power* (New Haven: Yale University Press, 1957). The concept of hydraulic civilizations is extremely important historically. The Marxian concept of oriental mode of production, which is linked to irrigation and thus to environment, was banned by the Soviet Communist party.

34. Neil Smith, *Uneven Development: Nature, Capital, and the Production of Space* (London: Basil Blackwell, 1984).

35. Karl Marx, *Capital* (New York: Vintage, 1967), 1:820, as quoted in Smith, *Uneven Development,* p. 64.

36. Karl Marx, *The Eighteenth Brumaire of Louis Bonaparte* (New York: International Publishers, 1963), p. 15.

37. Thomas Nagel, *The View from Nowhere* (New York: Oxford University Press, 1986), p. 90.

38. Ibid., pp. 90, 93.

39. As Nagel (ibid.) notes: "Realists always find it hard to say anything with which idealists cannot arrange to agree by giving it their own meaning" (p. 101).

Leszek Kolakowski, *Metaphysical Horror* (Oxford: Basil Blackwell, 1988), goes even farther in equating the two, for he believes that realism and idealism are interchangeable.

> If the presence of the observer cannot be removed from the description of some physical events, this does not necessarily imply that the observer is a Kantian intellect that imposes a-priori forms onto the shapeless stuff of perception: it is rather an intellect that discovers its own patterns in the reality as it verily is, and is able to reveal them because the reality is mind-like, and the very act of cognition, as Plato would have it, presupposes an affinity, or even a loving kinship, between my mind and the mind of the world. (p. 71)

40. Giddens, *The Constitution of Society,* p. 348. Derek Gregory was among the first to introduce the concept of reflexivity to geography; see his *Ideology, Science and Human Geography* (London: Hutchinson, 1978), esp. pp. 123–46.

41. See, for example, Talcott Parsons, ed., *Max Weber: The Theory of Social and Economic Organization* (New York: Free Press, 1947).

42. W. G. Runciman, *A Treatise on Social Theory,* vol. 1 (Cambridge: Cambridge University Press, 1983), believes this complication can be overcome. Still, it is especially complex when the experience of everyday life is molded by mass consumption and its disorienting qualities.

43. These meanings correspond to the meanings generally accepted in the natural sciences. Unfortunately, others have used the terms differently. For example David Harvey, *Social Justice and the City* (Baltimore: Johns Hopkins University Press, 1973), refers to absolute space as meaning either a fixed place or parcels of land that can be owned privately (or absolutely?); relative space is movement and connectivity. See J. Nicholas Entrikin, *The Betweenness of Place: Toward a Geography of Modernity* (Baltimore: Johns Hopkins University Press, 1990), for a critical analysis of Harvey's interpretation.

44. A. d'Abro, *The Rise of the New Physics* (New York: Dover, 1951), 1:112–13; and Albert Einstein's foreword to Max Jammer, *Concepts of Space* (New York: Harper and Brothers, 1969).

45. Jammer, *Concepts of Space,* p. xiv., refers to Einstein's comments.

46. "Although matter provides the epistemological basis for the metrical field, the fact must not be held to confer ontological primacy on matter over the field: matter is merely part of the field rather than its source," Adolph Grünbaum, *Philosophical Problems of Space and Time* (New York: Knopf, 1963), p. 421 and note 9, referring to Einstein's views. For another discussion of the current status of the view of absolute space, see Graham Nerlich, *The Shape of Space* (New York: Cambridge University Press, 1976).

47. Robert Sack, *Conceptions of Space in Social Thought: A Geographic Perspective* (Minneapolis: University of Minnesota Press, 1980), pp. 59–85.

48. See the works of Ernst Cassirer, especially *The Philosophy of Symbolic Forms,* 3 vols. (New Haven: Yale University Press, 1968); and those of Suzanne Langer, especially *Philosophy in a New Key: A Study in the Symbolism of Rea-*

son, Rite, and Art (Cambridge: Harvard University Press, 1942); *Feeling and Form* (New York: Scribner's, 1953); and *Mind: An Essay on Human Feeling,* 3 vols. (Baltimore: Johns Hopkins University Press, 1967).

49. Frederick Copleston, *A History of Philosophy* (New York: Doubleday, 1963), 6:238. Note that Kant's emphasis on Euclidean space can be abandoned without weakening his assumptions about the necessity of space to thought and awareness.

50. Quoted in ibid., p. 274.

51. Space applies more to our intuition of the external world ("space is the form of all appearances of the external world") and time to our internal world ("the form of the internal sense, or the intuition of ourselves"); ibid., p. 239. But since awareness is both internal and external, this distinction does not hold for long. Yet it has been the source of some lopsided philosophies, such as Henri Bergson's.

52. Ibid., p. 241.

53. See Cassirer, *Philosophy of Symbolic Forms;* and Langer, *Philosophy in a New Key;* Langer, *Feeling and Form;* Langer, *Mind.*

54. Sack, *Conceptions of Space in Social Thought.*

55. Thus distance or location is part of the process. And social geography, with its branches of economic, urban, political, and so forth, draws attention to how distance (in the relational form) affects human interactions and processes, especially by incorporating it within theories as a cost or friction that should be minimized.

56. Robert Sack, *Human Territoriality: Its Theory and History* (Cambridge: Cambridge University Press, 1986).

57. See, for example, Giddens, *Constitution of Society,* on distanciation; and David Harvey, *The Limits to Capital* (Chicago: University of Chicago Press, 1982), on the embedding of the built environment and the spatial fix within Marxist theory. Spatiality, according to Edward Soja, "The Spatiality of Social Life: Towards a Transformative Retheorisation," in Derek Gregory and John Urry, *Social Relations and Spatial Structures* (New York: Macmillan, 1985), is the assertion that space is simultaneously "a social product and an integral part of the material constitution and structuration of social life," both a "medium" and an "outcome" of social relations (pp. 92, 94). It is "socially produced space [and] must thus be distinguished from the physical space of material nature and the mental space of cognition and representation, each of which is used and incorporated into the social construction of spatiality but cannot be conceptualised as its equivalent" (pp. 92–93). See similar remarks in Edward Soja, *Postmodern Geographies: The Reassertion of Space in Critical Social Theory* (London: Verso, 1989), p. 120.

The central issue is, what distinguishes spatiality and how does it describe the role of space in social relations? The answer turns out to be that spatiality is primarily a rhetorical device to claim (not demonstrate) the importance of space. It is hardly an ontological position, for it does not show how space operates at the most basic levels in general social relations. Rather, its only specific example is a restatement of the role of space in capitalism, especially in the role of geographical inequality and in the spatial fix. And even in repeating this position, spatiality falls short of demonstrating the necessity of spatial inequalities in the social relations of capitalism. See Soja, *Postmodern Geographies,* pp. 107, 113. For example, it is conceded that, as a special case, capitalism can exist without significant areal variation of inequality (ibid., p. 113). And, even if spatial inequality were necessary for capitalism, this spatial unevenness still presupposes the existence of stable geographical territories and the effects of agglomeration and a relational distance decay. That is, spatiality assumes the existence of these more basic spatial principles while offering no new alternatives. Rather, it becomes yet another way of saying, but not showing, that our social actions are both constitutive of space and constituted by space.

58. See Harvey, *Limits to Capital,* and such general reviews of theory in geography as Peter Haggett, Andrew Cliff, and Allan Frey, *Locational Analysis in Human Geography,* 2d ed. (London: Edward Arnold, 1977).

· 5 ·
Place and Modern Culture

1. The "new" cultural geography discusses culture in the Gramscian terms of hegemony and struggle. Still, culture is also cooperation, and the primary contribution of cultural geography has been the understanding of how people build a world. See Peter Jackson, "Social Geography: Social Struggles and Spatial Strategies," *Progress in Human Geography* (1988): 263–69; and Steven Daniels, "Marxism, Culture, and the Duplicity of Landscape," in Nigel Thrift and Richard Peet, eds., *New Models in Geography: The Political-Economy Perspective* (London: Unwin-Hyman, 1989).

2. Clifford Geertz, "Thick Description," in Clifford Geertz, ed., *The Interpretation of Cultures* (New York: Basic Books, 1973).

3. Émile Durkheim, *The Division of Labor in Society* (Basingstoke, Hants.: Macmillan, 1933); Ferdinand Tonnies, *Community and Society* (East Lansing: Michigan State University Press, 1957).

4. Claude Lévi-Strauss, *The Savage Mind* (Chicago: University of Chicago Press, 1966).

5. Yi-Fu Tuan, *Segmented Worlds and Self: Group Life and Individual Consciousness* (Minneapolis: University of Minnesota Press, 1982), p. 177.

6. For a discussion of these utopias, see Charles Andrews, ed., *Famous Utopias* (New York: Tudor, n.d.); and Austin Wright, *Islandia* (New York: Farrar and Rinehart, 1942). I mention this last book because of its intricate geography and because it was written by the brother of the famous geographer John Kirtland Wright.

7. Philip Porter and Fred Lukermann, "The Geography of Utopia," in David Lowenthal and Martin Bowden, with Mary Lamberty, eds., *Geographies of the*

Mind: Essays in Historical Geography in Honor of John Kirtland Wright (New York: Oxford University Press, 1975), pp. 203, 206.

8. Tuan, *Segmented Worlds,* p. 177.

9. Ibid., p. 178.

10. Plato, *The Republic,* Jowett translation, bk. 3, line 414.

11. Robert Sack, *Conceptions of Space in Social Thought: A Geographic Perspective* (Minneapolis: University of Minnesota Press, 1980), p. 187. See also Yi-Fu Tuan, "The Significance of the Artifact," *Geographical Review* 70 (1980): 462–72, esp. p. 463.

12. Paul Wheatley, *The Pivot of the Four Quarters: A Preliminary Enquiry into the Origins and Character of the Ancient Chinese City* (Chicago: Aldine, 1971).

13. Eugene Walter, *Placeways: A Theory of the Human Environment* (Chapel Hill: University of North Carolina Press, 1988).

14. The assumption that these historical sites, memorials, and museums are really part of the past occurs not only in brochures provided to the visitors but in proposals to granting agencies. A grant proposal for Old World Wisconsin claims that the museum "truly represents the origins of most citizens of Wisconsin." I am indebted to Drew Ross for this and other information on the advertising of tourist places.

15. Sally Moore and Barbara Myerhoff, eds., *Secular Ritual* (Assen, The Netherlands: Van Gorcum, 1977). *Civil religion* was first used by Jean-Jacques Rousseau in *The Social Contract,* bk. 4, chap. 8, to include a belief in the "existence of God, the life to come, the reward of virtue and the punishment of vice, and exclusion of religious intolerance." Paraphrased in Robert Bellah, "Civil Religion in America," *Daedalus* 96 (1967): 1–21.

16. Bellah, "Civil Religion," p. 18.

17. Ibid.

18. Ken Foote, "Stigmata of National Identity: Exploring the Cosmography of America's Civil Religion," in M. Hurst and S. Wong, eds., *Person, Place, Things: Essays in Honor of Philip Wagner,* forthcoming.

19. Bellah, "Civil Religion," p. 7, claims that in the American case it has more substance. This echoes Durkheim's claim that religion has social roots in that it is based on belonging to a group or unit that transcends the immediate and local community.

20. American Indians are also immigrants. The only people who did not arrive on the continent voluntarily are African slaves.

21. Robert Sack, *Human Territoriality: Its Theory and History* (Cambridge: Cambridge University Press, 1986), p. 149.

22. James Mayo, *War Memorials as Political Landscape: The American Experience and Beyond* (New York: Praeger, 1988), p. 13. See also Wilbur Zelinsky, *Nation into State: The Shifting Symbolic Foundations of American Nationalism* (Chapel Hill: University of North Carolina Press, 1988), for a discussion of the numerous means by which American nationalism is made visible on the landscape.

23. Ibid., p. 19.

24. Ibid.

25. In earlier times, memorials were taken more seriously. In fact, buildings were usually dedicated and had at least a cornerstone. Now, the power of symbolism in general has diminished. "Parades are all but dead. It has become a popular form of theater with historical glamour but no history. It charms with entertainment that lacks commitment." J. B. Jackson, *Discovering the Vernacular Landscape* (New Haven: Yale University Press, 1984), p. 5.

26. See, for example, A. R. Zito, "City Gods, Filiality, and Hegemony in Late Imperial China," *Modern China* 13 (1987): 333–71.

27. See J. Nicholas Entrikin, *The Betweenness of Place: Toward a Geography of Modernity* (Baltimore: Johns Hopkins University Press, 1990); and J. Nicholas Entrikin, "Place, Region and Modernity," in John Agnew and James Duncan, eds., *The Power of Place: Bringing Together Geographical and Sociological Imaginations* (London: Unwin-Hyman, 1989), pp. 30–43.

28. This is the approach taken by Paul Adams, "Television and the Modern Experience of Space and Place," Master's thesis, University of Wisconsin—Madison, 1989. I am indebted to Adams for his insights and for drawing my attention to the following literature on television and context.

29. See Tony Schwartz, *Media: The Second God* (New York: Random House, 1981), p. 44; Neil Postman, *Amusing Ourselves to Death: Public Discourse in the Age of Show Business* (New York: Viking/Penguin, 1985), p. 38; Gregor Goethals, *The TV Ritual: Worship at the Video Altar* (Boston: Beacon, 1981), p. 142.

30. Edmund Carpenter, *Oh, What a Blow That Phantom Gave Me!* (New York: Holt, Rinehart, Winston, 1973), p. 63.

31. Marshall McLuhan, *The Medium Is the Massage* (New York: Bantam, 1967), p. 16.

32. Gary Gumpert, *Talking Tombstones and Other Tales of the Media Age* (New York: Oxford University Press, 1987), p. 9.

33. Ibid., p. 180.

34. Alan Rubin, Elizabeth Perse, and Donald Taylor, "A Methodological Examination of Cultivation," *Communication Research* 15 (1988): 107–34; quotation on p. 108.

35. Edward Relph, *Place and Placelessness* (London: Pion, 1976), pp. 58, 92.

36. See McLuhan, *The Medium Is the Massage;* and Adams, "Television and the Modern Experience," p. 16, who summarizes the position of John Fiske and John Hartley, *Reading Television* (London: Methuen, 1978), chap. 6, that television is like a "bard" whose words we decipher like a text.

37. Martyn Youngs, "The English Television Landscape Documentary: A Look at Granada," in Jacquelin Burgess and John Gold, eds., *Geography, the Media, and Popular Culture* (New York: St. Martin's, 1985). See also Jacquelin Burgess, "Landscapes in the Living Room: Television and Landscape Research," *Landscape Research* 12 (1987): 1–7.

38. A case in point is that

> in 1989 American television was censured for simulating a scene in which a purported spy passed a briefcase to a Soviet agent: such "fakery" was held to insult viewers' views, ethics, and journalism. Network executives defended the simulated episode on the grounds that old distinctions between truth and fiction were passé. After all, many Americans considered their favorite TV character more 'real' than their friends and neighbours.

David Lowenthal, "Forging the Past," typescript, quoting Charles Bremner, *International Herald Tribune,* 29 July 1989; and the *Times,* 9 August 1989. Television news also fits events to conform to its narrative conventions. See James Carey, ed., *Media, Myths, and Narratives: Television and the Press* (Beverly Hills: Sage, 1988).

39. Joshua Meyrowitz, *No Sense of Place: The Impact of Electronic Media on Social Behavior* (New York: Oxford University Press, 1985), pp. 6, 214, 235, 323.

40. Ibid., p. 176. The quotations in this paragraph are from ibid., pp. 309, 149, and 176.

41. Ibid., p. 176.

42. Gavriel Salomon, "The Study of Television in a Cross-Cultural Context," *Journal of Cross-Cultural Psychology* (*Special Issue: Television in the Developing World*), 16 (1985): 381–97.

43. The flow of images can be disorienting and make it appear as though there is no true referent. Television becomes the "generation of models of a real without origin or reality: a hyperreal." Jean Baudrillard, *Simulations* (New York: Semiotext[e], 1983), p. 2. Chen echoes the same feelings: "images and codes, the subject . . . and the event . . . flow into each other, intersect with each other, and refer to each other without worrying about the 'true' referent." Kuan-Hsing Chen, "The Masses and the Media: Baudrillard's Implosive Postmodernism," *Theory, Culture and Society* 4 (1987): 71–88; quotation on pp. 71–72.

· 6 ·
A Geographical Model of Consumption

1. Daniel Boorstin, *The Americans: The Democratic Experience* (New York: Random House, 1973); and Henri Lefebvre, *Everyday Life in the Modern World,* trans. S. Rabinovitch (New Brunswick, N.J.: Transaction Books, 1984). Lefebvre emphasizes the fact that our consumption is planned and coordinated by industry and the state.

2. Consumption forms a mass culture. T. Jackson Lears, "From Salvation to Self-Realization: Advertising and the Therapeutic Roots of the Consumer Culture 1880–1930," in R. W. Fox and T. J. Lears, eds., *The Culture of Consumption: Critical Essays in American History, 1880–1930* (New York: Pantheon, 1983). Boorstin, *The Americans;* Lefebvre, *Everyday Life.* Also note that the separation of production from consumption is not new (see Michael Chisholm, "The Increasing Separation of Production and Consumption," in B. L. Turner et al., eds.,

The Earth as Transformed by Human Action [Cambridge: Cambridge University Press, 1990]). What is new is the quality of the separation.

3. People in even the remotest parts of the Third World are influenced by advertising, to the point where peasant subsistent farmers enter some aspects of the commerical economy in order to buy consumer durables, often radios. These then expose them to further advertisements. In addition, the Third World landscape contains billboards and other forms of advertising, which in some cases are more obtrusive than in the United States.

4. Mary Douglas and Baron Isherwood, *The World of Goods* (New York: Basic Books, 1979).

5. Hannah Arendt, *The Human Condition: A Study of the Central Dilemmas Facing Modern Man* (Chicago: University of Chicago Press, 1958), p. 137.

6. Karl Marx, "The Manifesto of the Communist Party," in Karl Marx and Friedrich Engels, *Selected Works* (Moscow: Foreign Languages Publishing House, 1955), 1:37.

7. Lefebvre, *Everyday Life,* p. vii.

8. This model draws on Robert Sack, "The Consumer's World: Place as Context," *Annals of the Association of American Geographers* 78 (1988): 642–64.

9. This message is identified in ibid. and stems from a geographical understanding of the role of consumption. Other definitions of advertising, or its function, are noteworthy. For example, Lefebvre, *Everyday Life,* claims: "Advertising is the poetry of the modern world: and what it promises is freedom from fear. . . . Ads address terror from alienation," p. xv. R. Belk and R. Pollay call them "images of ourselves"; see their "Images of Ourselves: The Good Life in 20th Century Advertising," *Journal of Consumer Research* 11 (1989): 53–63. Lears, "From Salvation to Self-Realization," pp. 3–38, says ads unintentionally reinforce and spread a culture of consumption in which the presence of material goods has replaced the pursuit of religious goods. In "Some Versions of Fantasy: Toward a Cultural History of American Advertising 1880–1930," *Prospects* 9 (1979): 349–405, Lears claims that "even the magic is without purpose. It is not placing us within a universal context," p. 398. Christopher Lasch, *The Culture of Narcissism: American Life in an Age of Diminishing Expectations* (New York: Norton, 1979), contends that advertising has promoted consumption as the principal value in modern life. And Boorstin sees advertising as fostering a community of consumption in which the commodity is the principal device that binds our culture together; see his *The Americans.* All of these point to the power of advertising to make consumption essential to shared experiences. They do not, however, claim that ads promote particular products as much as they promote the entire enterprise of consuming. This is what Michael Schudson makes explicit in his *Advertising, the Uneasy Persuasion: Its Dubious Impact on American Society* (New York: Basic Books, 1984).

10. See Schudson, *Advertising,* for an excellent review of the literature on the effectiveness of ads. The points in this paragraph come from this source.

11. Because ads are both in and above the realm of meaning, they act like a perspective, but one that is barely self-critical.

12. For an excellent history of our relationship to nature and our various conceptions of it, see Clarence Glacken, *Traces on the Rhodian Shore: Nature and Culture in Western Thought from Ancient Times to the End of the Eighteenth Century* (Berkeley and Los Angeles: University of California Press, 1967). See also R. G. Collingwood, *The Idea of Nature* (Oxford: Clarendon, 1945).

13. P. F. Strawson, *Individuals: An Essay in Descriptive Metaphysics* (New York: Anchor/Doubleday, 1963), argues that the most natural system for individuation is a spatial one. See also the more empirical observations of A. Michotte, *The Perception of Causality* (London: Methuen, 1963).

14. A lengthier discussion of this conceptual separation and recombination is found in Robert Sack, *Conceptions of Space in Social Thought: A Geographic Perspective* (Minneapolis: University of Minnesota Press, 1980).

15. Robert Sack, *Human Territoriality: Its Theory and History* (Cambridge: Cambridge University Press, 1986).

16. The expression "a world of strangers" is from Lynn Lofland, *A World of Strangers* (New York: Basic Books, 1973). For the same sentiments, see David Riesman, *The Lonely Crowd* (New Haven: Yale University Press, 1950); Richard Sennett, *The Fall of Public Man* (New York: Vintage, 1978); and Michael Ignatieff, *The Needs of Strangers* (New York: Penguin, 1984). The sense of the word *strangers* is complex, and its meaning has changed over time. Joshua Meyrowitz, "The Three Worlds of Strangers: Boundary Shifts and Changes in 'Them' vs. 'Us'," *Annals of the Association of American Geographers* 80 (1990): 129–32, identifies three stages in the meaning of the term, which correspond to changes in modes of communication; the present meaning of *stranger* connotes some familiarity. Anthony Giddens, *The Consequences of Modernity* (Stanford: Stanford University Press, 1990), analyzes how the modern meaning of *strangers* contains an element of trust. I agree with these points about this term, and yet I find it the best term for capturing contemporary relationships.

17. Meyrowitz, "The Three Worlds of Strangers"; Giddens, *The Consequences of Modernity.*

18. David Harvey, *Consciousness and the Urban Experience: Studies in the History and Theory of Capitalist Urbanization* (Baltimore: Johns Hopkins University Press, 1985), p. 5.

19. Erving Goffman, *The Presentation of Self in Everyday Life* (New York: Doubleday, 1959).

20. Jean Baudrillard, *Simulations* (New York: Semiotext[e], 1983). The idea that everything is now a play of signs is an extension of the Saussurian tradition.

21. Yi-Fu Tuan, *Segmented Worlds and Self: Group Life and Individual Consciousness* (Minneapolis: University of Minnesota Press, 1982).

22. Clifford Geertz, "Thick Description," in Clifford Geertz, ed., *The Interpretation of Cultures* (New York: Basic Books, 1973).

23. Attempts have been made to use anthropological structuralism to analyze contemporary culture. These often address small issues, such as sporting events or funeral rites, and do not capture the sweep of cultural meanings. The most impressive attempt has been Marshall Sahlins, *Culture and Practical Reason*

(Chicago: University of Chicago Press, 1976), chap. 4. One can argue that Marxist analysis of capitalism is akin to a structural approach. The dialectical relations between labor and capital are similar to the mediations and oppositions of structural anthropology.

24. A similar problem plagues all areas except science. In art, the symbolic products are constantly brought into question. See William Barrett, *Time of Need: Forms of Imagination in the Twentieth Century* (New York: Harper and Row, 1972); and Richard Schechner, *Essays on Performance Theory: 1970–1976* (New York: Drama Book Specialists, 1977).

25. Stephen Toulmin, "The Inwardness of Mental Life," *Critical Inquiry* 6 (1979–80): 1–16.

26. Tuan, *Segmented Worlds and Self.*

27. Alfred Sohn-Rethel, *Intellectual and Manual Labour: A Critique of Epistemology* (London: Macmillan, 1978), argues that capitalism and money create these distinctions.

28. Ibid., pp. 48–49.

29. Mark Poster generalizes this point:

> hence the unique contradiction of modernity: on the one hand, the progress of technology led to the organized socialization of all experiences; on the other, a new type of existence [in] daily life emerged and stagnated outside the general movement of history. People were at once organized into activities that were complex and controlled; yet in their activities they were totally unaware of themselves as social, public, beings. . . . What maintained the precarious balance between the masses' passivity and their dutiful, punctual, if apathetic performance of their functions was the role of new commodities and the mass media in their lives.

Mark Poster, *Existential Marxism in Postwar France* (Princeton: Princeton University Press, 1985), p. 246, as quoted in Barney Warf, "Ideology, Everyday Life and Emancipatory Phenomenology," *Antipode* 18 (1986): 268–83; quotation on 276; inserted word is Warf's.

30. See Schudson, *Advertising, the Uneasy Persuasion;* Roland Marchand, *Advertising the American Dream: Making Way for Modernity, 1920–1940* (Berkeley and Los Angeles: University of California Press, 1985); and S. Fox, *The Mirror Makers: A History of American Advertising and Its Creators* (New York: Morrow, 1984). For a general early history of the methods of advertising, see Frank Presbrey, *The History and Development of Advertising* (Garden City, N.Y.: Doubleday, 1929).

31. Marxists might say that commoditization obliterates social relations.

32. Labels may say where a product was assembled but not where all of the parts originated. Some labels that do include the latter are so general that it is impossible to know what to do with the information. Consider a label on a tool that states that parts were manufactured in Taiwan, South Korea, the Philippines, Hong Kong, and the United States.

33. Douglas and Isherwood, *The World of Goods.*

34. Excellent examples are Varda Leymore, *Hidden Myth: Structure & Symbolism in Advertising* (New York: Basic Books, 1975); Judith Williamson, *Decoding Advertisements: Ideology and Meaning in Advertising* (London: Marion Boyars, 1978); William Leiss, Stephen Kline, and Sut Jhally, *Social Communication in Advertising* (New York: Methuen, 1986); and Sut Jhally, *The Codes of Advertising: Fetishism and the Political Economy of Meaning in the Consumer Society* (New York: St. Martin's, 1987).

35. For a discussion of Marxian fetishism and commodities, see Sut Jhally, *The Codes of Advertising;* and Michael Taussig, *The Devil and Commodity Fetishism in South America* (Chapel Hill: University of North Carolina Press, 1980).

36. See Richard Pollay, ed., *Information Sources in Advertising History* (Westport, Conn.: Greenwood, 1979); Richard Pollay, "The Subsiding Sizzle: Shifting Strategies in Twentieth-Century Magazine Advertising," University of British Columbia, Faculty of Commerce, History of Advertising Archives, 1983; and Richard Pollay, "The Identification and Distribution of Values Manifest in Print Advertising, 1900–1980," University of British Columbia, Faculty of Commerce, History of Advertising Archives; and Leiss, Kline, and Jhally, *Social Communication in Advertising,* for historical trends in advertising.

37. T. Jackson Lears, *No Place of Grace: Antimodernism and the Transformation of American Culture, 1880–1920* (New York: Pantheon, 1981).

38. David Ward, *Cities and Immigrants: A Geography of Change in Nineteenth-Century America* (New York: Oxford University Press, 1971), p. 3.

39. Ibid., p. 87.

40. See the works cited in notes 30 and 34 for histories of advertising.

41. David Hounshell, *From the American System to Mass Production, 1800–1932* (Baltimore: Johns Hopkins University Press, 1984).

42. Pollay, "The Subsiding Sizzle"; and Leiss, Kline, and Jhally, *Social Communication in Advertising,* provide excellent discussions of the development of advertising.

43. See Stuart Ewen, *Captains of Consciousness: Advertising and the Social Roots of the Consumer Culture* (New York: McGraw-Hill, 1976), for a general analysis of the relation between social change and advertising.

44. Eli Zaretsky, *Capitalism, the Family, and Personal Life* (New York: Harper and Row, 1976); and M. Jay, *Adorno* (Cambridge: Harvard University Press, 1984).

45. Sigfried Giedion, *Mechanization Takes Command: A Contribution to Anonymous History* (New York: Oxford University Press, 1984), pp. 329–31.

46. See the works cited in notes 34 and 36.

47. Claude Lévi-Strauss, *The Raw and the Cooked* (New York: Harper and Row, 1970). See also the concept of second nature in Glacken, *Traces on the Rhodian Shore;* for a Marxist twist, see Neil Smith, *Uneven Development: Nature, Capital, and the Production of Space* (London: Basil Blackwell, 1984).

48. See, for example, Lefebvre, *Everyday Life.*

49. William Leiss, *The Limits to Satisfaction: An Essay on the Problems of Needs and Commodities* (Toronto: University of Toronto Press, 1976), p. 88;

Lasch, *The Culture of Narcissism,* pp. 15–19, 30–38; and Alasdair MacIntyre, *After Virtue: A Study in Moral Theory* (Notre Dame: University of Notre Dame Press, 1981), p. 31, who considers the philosophical consequences of this fragmentation.

50. Guy Debord, *Society of the Spectacle* (Detroit: Black and Red, 1973).

51. Williamson, *Decoding Advertisements;* Leiss, Kline, and Jhally, *Social Communication in Advertising.*

52. The example is from Williamson, *Decoding Advertisements.*

53. Ibid.

54. Raymond Williams, "Advertising: The Magic System," in his *Problems in Materialism and Culture* (London: New Left Books, 1978). Lears, "Some Versions of Fantasy," p. 398, argues that ads even talk about magic, but they have no coherent symbolic order and no higher moral purpose, and they do not place us within a universal context.

55. This model of magic or ritual as conflation of symbol and referent is developed in greater detail in Sack, *Conceptions of Space in Social Thought.*

56. Consider the route taken by David Harvey, "Flexible Accumulation through Urbanization: Reflections on 'Post Modernism' in the American City," *Antipode* 19 (1987): 260–86; and David Harvey, *The Condition of Postmodernity: An Enquiry into the Origins of Cultural Change* (Oxford: Basil Blackwell, 1989).

· 7 ·
Places of Consumption

1. In Paris, a journey by foot that takes fifteen minutes is estimated to have taken an hour and a half at the beginning of the nineteenth century. See Richard Sennett, *The Fall of Public Man* (New York: Vintage, 1976), p. 143. What is true for the ancien régime in Paris applies for most of eighteenth-century-to-mid-nineteenth-century Western Europe and North America.

2. Ibid., p. 141.

3. See Stuart Ewen and Elizabeth Ewen, *Channels of Desire: Mass Images and the Shaping of American Consciousness* (New York: McGraw-Hill, 1982), for a discussion of one of the earliest shifts to mass production in North America—that of the clothing industry.

4. Sennett, *The Fall of Public Man,* p. 144.

5. William Leach, "Transformations in a Culture of Consumption: Women and Department Stores, 1890–1925," *Journal of American History* 71 (1984): 322.

6. Sennett, *The Fall of Public Man,* p. 144.

7. Émile Zola, *Au Bonheur des Dames,* quoted in H. Pasdermadjian, *The Department Store* (London: Newman Books, 1954), p. 12. See also Sennett, *The Fall of Public Man.*

8. Émile Zola is quoted in Michael Miller, *The Bon Marché: Bourgeois Cul-*

ture and the Department Store, 1869–1920 (Princeton: Princeton University Press, 1981), p. 5; G. D'Avenel, "Les Commerces en Grands Magasins," *Revue des Deux Mondes* (1892): 136, quoted in Sennett, *The Fall of Public Man,* p. 144; and G. D'Avenel as quoted in Pasdermadjian, *The Department Store.*

9. Miller, *The Bon Marché,* p. 168.

10. Grant MacGracken, *Culture and Consumption: New Approaches to the Symbolic Character of Consumer Goods and Activities* (Bloomington: Indiana University Press, 1988), p. 24.

11. Sennett, *The Fall of Public Man,* p. 144.

12. Ibid., pp. 144–45.

13. Miller, *The Bon Marché,* pp. 202–3.

14. Quoted in Leach, "Transformations," p. 326.

15. Miller, *The Bon Marché;* and Leach, "Transformations."

16. Leach, "Transformations," p. 322.

17. Miller, *The Bon Marché,* pp. 184–85, 186.

18. Susan Benson, *Counter Cultures: Saleswomen, Managers, and Customers in American Department Stores, 1890–1940* (Urbana: University of Illinois Press, 1986).

19. Sennett, *The Fall of Public Man,* p. 142.

20. Miller, *The Bon Marché,* pp. 61–62.

21. Gordon Weil, *Sears, Roebuck, USA: The Great American Catalog Store and How It Grew* (Briarcliff Manor, N.Y.: Stein and Day, 1977).

22. Godfrey Lebhar, *Chain Stores in America: 1859–1950* (New York: Chain Store Publishing Corporation, 1952), p. 9.

23. Philip Langdon, *Orange Roofs, Golden Arches* (New York: Knopf, 1986), p. 30.

24. John Baeder, *Gas, Food, and Lodging* (New York: Abbeville, 1982).

25. Edward Relph, *The Modern Urban Landscape* (London: Croom Helm, 1987), p. 158.

26. Ibid., p. 83. This paragraph is from ibid., pp. 84, 182–88.

27. Tom Wolfe, *The Kandy-Kolored Tangerine-Flake Streamline Baby* (New York: Farrar, Straus, Giroux, 1965), p. 8.

28. Langdon, *Orange Roofs, Golden Arches,* p. 150.

29. Bruce Axler, *Foodservice: A Managerial Approach* (D. C. Heath for NIFI Textbook, 1979), p. 77.

30. Kenneth Foote, *Color in Public Spaces: Toward a Communication-Based Theory of the Urban Built Environment,* Research Paper 205 (Chicago: University of Chicago, Department of Geography, 1983).

31. Relph, *The Modern Urban Landscape,* pp. 168–70.

32. J. Ross McKeever and Nathaniel Griffin, with Frank Spink, Jr., *Shopping Center Development Handbook* (Washington: Urban Land Institute, 1977), p. 1931.

33. William Kowinski, *The Malling of America: An Inside Look at the Great Consumer Paradise* (New York: Morrow, 1985), p. 57.

34. McKeever and Griffin, *Shopping Center Development,* p. 72.

35. George Sternbieb, ed., *Shopping Centers U.S.A.* (New York: Center for Urban Policy Research, 1981), p. 241.

36. McKeever and Griffin, *Shopping Center Development,* p. 96.

37. Kowinski, *The Malling of America,* p. 22.

38. Ibid., p. 377.

39. McKeever and Griffin, *Shopping Center Development,* p. 240.

40. Kowinski, *The Malling of America,* p. 20.

41. Peter Hall, *Cities of Tomorrow: An Intellectual History of Urban Town Planning and Design in the Twentieth Century* (Cambridge: Basil Blackwell, 1988), p. 348.

42. Ibid., pp. 350–51.

43. N. Falk, "Baltimore and Lowell: Two American Approaches," *Built Environment* 12 (1986): 145–52.

44. Hall, *Cities of Tomorrow,* p. 351.

45. Yi-Fu Tuan, *Segmented Worlds and Self: Group Life and Individual Consciousness* (Minneapolis: University of Minnesota Press, 1982).

46. Sennett, *The Fall of Public Man;* the internal partitioning of the house is discussed in chapter 2. See also Tuan, *Segmented Worlds and Self.*

47. Gaston Bachelard, *The Poetics of Space* (Boston: Beacon, 1964), makes the association transhistorical and phenomenological. See also Peter Wilson, *The Domestication of the Human Species* (New Haven: Yale University Press, 1988), for a fascinating discussion of the role of the dwelling in human development. The walls of our homes anchor us to place and introduce the possibility of hidden actions and private lives not possible in predomesticated society.

48. R. M. Rackoff, "Ideology in Everyday Life: The Meaning of the House," *Politics and Society* 7 (1977): 85–104, esp. p. 94.

49. S. Weir Mitchell, "When College is Hurtful to the Girl," *Ladies Home Journal* 17 (1900): 14; this article was drawn to my attention by Donna Baron.

50. Gerry Pratt, "The House as an Expression of Social Worlds," in James Duncan, ed., *Housing Identity: Cross-Cultural Perspectives* (London: Croom Helm, 1981).

51. Billy Baldwin, *Billy Baldwin Decorates* (New York: Holt, Rinehart, Winston, 1972), pp. 38 and 168.

52. Judith Goldstein, "Lifestyles of the Rich and Tyrannical," *American Scholar* (Spring 1987): 235–47.

53. *New York Times,* 1 March 1986, quoted in ibid., p. 240.

54. See William Leiss, *The Limits to Satisfaction: An Essay on the Problem of Needs and Commodities* (Toronto: University of Toronto Press, 1976).

55. O. S. Fowler, *A Home for All* (New York: Fowler and Wills, 1854).

56. David Handlin, *The American Home, Architecture and Society, 1815–1915* (Boston: Little, Brown, 1974).

57. Sears and Roebuck, *Homes of Today* (n.p.: Sears, Roebuck, and Co., 1931).

58. Edward Soja, *Postmodern Geographies: The Reassertion of Space in Critical Social Theory* (London: Verso, 1989), p. 231.

59. The precursor to condominiums was called a planned unit development. See Marshall Block and W. Ingersoll, eds., *Timesharing* (Washington, D.C.: Urban Land Institute), pp. 4–5. See also *Condominiums, Housing for Tomorrow* (Boston: Management Report, 1984).

60. The Marxian conception of alienation is based on the alienation of the means of production and the extraction of surplus value. But the same effects could result simply by the technical division of labor, hierarchy, and bureaucracy. In this respect, Weber rather than Marx might have more of lasting value to say.

61. These issues are discussed at length in Robert Sack, *Human Territoriality: Its Theory and History* (Cambridge: Cambridge University Press, 1986), esp. chap. 6.

62. Paul Fussell, "The Stationary Tourist: Around the World Going Nowhere," *Harpers* (April 1979): 31–38; quotation on pp. 31–32.

63. Daniel Boorstin, *The Image: A Guide to Pseudo-Events in America* (New York: Atheneum, 1961), p. 80.

64. Judith Adler, "Origins of Sightseeing," *Annals of Tourism Research* 16 (1989): 7–29; quotation on p. 11.

65. Selling the idea of losing oneself in order to discover oneself is common advertising practice.

66. Walter Christaller, "Some Considerations of Tourism Location in Europe," *Regional Science Association Papers* 12 (1964): 95–105; quotation on p. 103.

67. R. W. Butler, "The Concept of a Tourist Area Cycle of Evolution: Implications for Management of Resources," *Canadian Geographer* 24 (1980): 5–12. The quotations and examples that follow are from ibid., pp. 7, 8, and 10.

68. A perfect case is the erosion of the beach in Grand Isle, Louisiana. Klaus J. Meyer-Arendt, "The Grand Isle, Louisiana Resort Cycle," *Annals of Tourism Research* 12 (1985): 449–65.

69. Tourism transforms not only the natural environment but also the social environment—to the point where the indigenous people have trouble surviving and the reason for the tourist trade is defeated. Gary Hovinen, "Visitor Cycles: Outlook for Tourism in Lancaster County," *Annals of Tourism Research* 9 (1982): 565–83, shows stresses that tourism places on the Amish community in Lancaster, Pennsylvania. The Amish are the county's chief tourist attraction: the county has been described as among the top ten tourist centers in the United States. For excellent studies on the complex effects of tourism, see Stephen Britton and William Clarke, eds., *Ambiguous Alternatives: Tourism in Small Developing Countries* (Suva, Fiji: University of the South Pacific, 1987); and Ken Olwig and Karen Olwig, "Underdevelopment and the Development of National Park Ideology," *Antipode* 6 (1974): 16–26.

70. Erik Cohen, "Authenticity and Commoditization in Tourism," *Annals of Tourism Research* 15 (1988): 371–86; quotation on p. 381. Cohen's analysis is based on D. J. Greenwood, "Culture by the Pound: An Anthropological Perspective on Tourism as Cultural Commoditization," in V. L. Smith, ed., *Hosts and Guests: The Anthropology of Tourism* (Philadelphia: University of Pennsylvania Press, 1977), pp. 129–39.

71. Greenwood, "Culture by the Pound," p. 135, quoted in Cohen, "Authenticity and Commoditization," p. 381.

72. Cohen, "Authenticity and Commoditization," p. 382, based on P. F. McKean, "Tourism, Culture Change, and Culture Conservation in Bali," in D. J. Banks, ed., *Changing Identities in Modern Southeast Asia* (The Hague: Mouton, 1976), pp. 237–48.

73. McKean, "Tourism, Culture Change," p. 244, quoted in Cohen, "Authenticity and Commoditization," p. 382.

74. Boorstin, *The Image.*

75. Ibid., p. 99.

76. Cohen, "Authenticity and Commoditization," p. 372.

77. Erik Cohen, "'Primitive and Remote': Hill Tribe Trekking in Thailand," *Annals of Tourism Research* 16 (1989): 30–61.

78. Cohen, "Authenticity and Commoditization," p. 32.

79. Ibid., p. 56.

80. See Donald Redfoot, "Tourist Authenticity, Touristic Angst, and Modern Reality," *Qualitative Sociology* 7 (1984): 291–309, for tourists' assumptions about authenticity.

81. Dean MacCannell, *The Tourist: A New Theory of the Leisure Class* (New York: Schocken, 1975), p. 13.

82. Cohen, "Authenticity and Commoditization," p. 380.

83. Umberto Eco, *Travels in Hyperreality* (New York: Harcourt, Brace, Jovanovich, 1983), p. 8.

84. Ibid., paraphrased in Drew Ross, "Tourism and the Modern World," Department of Geography, University of Wisconsin–Madison (1989), pp. 9–10.

85. These issues are addressed by David Lowenthal in several of his publications; see especially his *The Past Is a Foreign Country* (New York: Cambridge: Cambridge University Press, 1985).

86. Eco, *Travels in Hyperreality,* pp. 7, 15.

87. Jean Baudrillard, *Simulations* (New York: Semiotext[e], 1983), p. 15.

88. It is ironic that at the peak of this consumer disorientation, remote sensing and satellite imagery provide us with accurate knowledge of the location of things on the earth; we possess the power to manipulate this information with great accuracy and ease in geographical information systems. But this knowledge is now becoming so specialized and segmented that it even threatens the unity of geographical knowledge.

89. Margaret King, "Disneyland and Walt Disney World: Traditional Values in Futuristic Form," *Journal of Popular Culture* 15 (1981): 116–40; quotation on p. 117.

90. Ibid., p. 121.

91. Walt Disney World Vacation Kingdom, *Disney University Teaches People Philosophy.*

92. Walt Disney Vacation Kingdom, college program, An Introduction to Walt Disney World Entertainment.

93. *Harpers,* August 1983.

94. The catalogue describes a replica of Sleeping Beauty's castle in these words: "Who could ever forget the magic of Sleeping Beauty's Castle? And now the splendor of this fairy tale palace is captured in this bronzed sculpture castle. The exquisite detail and finely sculptured towers evoke the romance and enchantment of childhood dreams and fairy tales." *Walt Disney World Merchandise Catalogue,* p. 10.

95. Ibid., p. 20.

96. Hidetoshi Kato, "Gilded Expectations: An Aspect of Intercultural Communication." (n.p.: East-West Center Communication Institute, 15 June 1973), p. 1. Speech given to Club 15. Quoted in King, "Disneyland and Walt Disney World," p. 128.

97. William Thompson, "Looking for History in L.A.," in his *At the Edge of History* (New York: Harper and Row, 1971), p. 21. Quoted in King, "Disneyland and Walt Disney World," p. 128.

98. Debora Silverman, *Selling Culture: Bloomingdale's, Diana Vreeland, and the New Aristocracy of Taste in Reagan's America* (New York: Pantheon, 1986). The following is based on this book; see pp. 21, 22, 24, 25, 29, 31–34.

99. Cohen, "Authenticity and Commoditization," p. 374.

100. Baudrillard, *Simulations,* p. 25.

101. Ibid., p. 11.

102. Michel Foucault, *The Order of Things* (New York: Vintage, 1970); and Robert Sack, *Conceptions of Space in Social Thought: A Geographic Perspective* (Minneapolis: University of Minnesota Press, 1980), chap. 6. See also Robert Sack, "Magic and Space," *Annals of the Association of American Geographers* 66 (1976): 309–22; and Denis Cosgrove, "Environmental Thought and Action: Premodern and Postmodern," *Transactions, Institute of British Geographers* 15 (1990): 344–58, who argues that postmodernism is similar in its logic to the Renaissance view of symbols.

103. Drew Ross, "Placing Authenticity: Mind and Environment in a Leisure Landscape, Wisconsin Dells, Wisconsin," Master's thesis, University of Wisconsin—Madison, 1989, discusses Heidegger and authenticity.

104. See Hannah Arendt, *The Human Condition: A Study of the Central Dilemmas Facing Modern Man* (Garden City, N.Y.: Doubleday, 1959), for a discussion of this inversion. Also see Lionel Trilling, *Sincerity and Authenticity* (Cambridge: Harvard University Press, 1972).

105. The same idea is echoed by Peter Berger:

> If nothing on "the outside" can be relied upon to give weight to the individual's sense of reality, he is left no option but to burrow into himself in search of the real. Whatever this sense of realism may then turn out to be, it must necessarily be in opposition to any external social formation. The opposition between self and society has now reached its maximum. The concept of authenticity is one way of articulating this experience.

Peter Berger, "'Sincerity and Authenticity' in Modern Society," *Public Interest* (1973): 81–90; quotation on p. 88.

106. Cohen, "Authenticity and Commoditization."

107. Christopher Lasch, *The Culture of Narcissism: American Life in an Age of Diminishing Expectations* (New York: Norton, 1979).

108. These thoughts are drawn from Cohen, "Authenticity and Commoditization."

109. Kimberly Dovey, "The Quest for Authenticity and the Replication of Environmental Meaning," in David Seamon and Robert Mugerauer, eds., *Dwelling, Places and Environment: Towards a Phenomenology of Person and World* (The Hague: Martinus Nijhoff, 1985), pp. 44–45.

· 8 ·
Place, Morality, and Consumption

1. This point is made by Tuan in his pioneering works on geography and morality. See Yi-Fu Tuan, *The Good Life* (Madison: University of Wisconsin Press, 1986); and Yi-Fu Tuan, *Morality and Imagination: Paradoxes of Progress* (Madison: University of Wisconsin Press, 1989).

2. See, for example, David Smith, *Geography, Inequality, and Society* (Cambridge: Cambridge University Press, 1987).

3. The question then becomes, How does the divine power enter our lives? Here again, the issue of the realms emerges as the particular paths through which the sense of this ineffable good is transmitted.

4. Tuan, *The Good Life,* p. 157.

5. This paragraph is based on ibid., pp. 114, 9–10.

6. Ibid., pp. 9–10.

7. Campbell traces consumerism back to romanticism and to strands in the Protestant ethic. See Colin Campbell, *The Romantic Ethic and the Spirit of Modern Consumerism* (Oxford: Basil Blackwell, 1987).

8. Norbert Elias, *The History of Manners* (New York: Pantheon, 1978), 1:79.

9. Ibid., p. 232.

10. Rosalind Williams, *Dream Worlds: Mass Consumption in Late Nineteenth-Century France* (Berkeley and Los Angeles: University of California Press, 1982), p. 24.

11. Ibid., p. 37.

12. Ibid., p. 38.

13. These arguments are part of the cultural heritage of even socialist countries, which presents an interesting moral dilemma. These goods were produced and accumulated by exploitation, and yet they became the artifacts that cultures point to as their highest achievements.

14. Williams, *Dream Worlds,* pp. 4, 7.

15. See, for example, criticisms of philosophy by Alasdair MacIntyre, *After Virtue: A Study in Moral Theory* (Notre Dame: University of Notre Dame Press, 1981); Bernard Williams, *Ethics and the Limits of Philosophy* (Cambridge: Harvard University Press, 1985); Thomas Nagel, *A View from Nowhere* (New York:

Oxford University Press, 1986); Richard Norman, *The Moral Philosophers* (Oxford: Oxford University Press, 1983); Jeffrey Stout, *Ethics after Babel* (Boston: Beacon, 1988); and Michael Walzer, *Spheres of Justice* (New York: Basic Books, 1985). All argue for a moral theory embedded in social context. In this sense, they are Aristotelian. Norman (pp. 188, 240) claims that one cannot have a morality focused on a nonsocial, isolated self; rather, morality must be based on human needs and social relations.

Williams makes a distinction between moral and theoretical positions:

> In part, it is because the scientific understanding of the world is not entirely consistent with recognizing that we occupy no special position in it but also incorporate, now, that recognition. The aim of ethical thought, however, is to help us to construct a world that will be our world in which we have a social, cultural and personal life. (p. 111)

And Stout calls the result of the view from nowhere "moral esperantoism." While they all look to local context for a better understanding of morality, they do not formally consider the role of place.

16. Technically, a pure public good needs to possess three characteristics. The first is joint supply, which means that the supply of a particular quantity to any one person does not diminish the possibility of supplying the same quantity of the good at the same price to any and all others. The second characteristic is the impossibility of exclusion (nonexcludability). This characteristic has two parts: (a) the supply to any one person prevents the good being withheld from any other person wishing access to it; (b) people who do not pay for the good or service cannot be excluded from its benefits, or the free-rider problem. The other side of the free-rider problem is that people tend to underreveal their preferences for this kind of good. The third characteristic is the impossibility of rejection (nonrejectability): this means that, once a service is supplied, it must be fully and equally consumed by all, even those who might not wish to do so. A public good is impure to the extent that one or all of these characteristics are not held. This definition, by Robert Sack, *Human Territoriality: Its Theory and History* (Cambridge: Cambridge University Press, 1986), p. 157, is a composite from Robert Bennett, *The Geography of Public Finance* (London: Methuen, 1980), pp. 11–13; and Paul Samuelson, "The Pure Theory of Public Expenditure," *Review of Economics and Statistics* 36 (1954): 387–89.

17. Of course, the index can be refined so that it combines geographical accessibility to hospitals with such factors as ability to pay, time to get to the hospital, and education of hospital users. These combined factors can then be mapped.

18. Richard Hartshorne, *The Nature of Geography: A Critical Survey of Current Thought in the Light of the Past* (Lancaster, Pa.: Association of American Geographers, 1939; reprinted 1961), believes that the geographer's task is to describe and explain areal variation over the earth. "Geography . . . can demand serious attention if it strives to provide complete, accurate, and organized knowledge to satisfy man's curiosity about how things differ in the different parts of the world" (p. 131). Geography is thus concerned with studying the areal variation of the world.

19. Sack, *Human Territoriality,* p. 36.

20. As far as I can tell, no one else has ever made the point that territoriality is important to equal access. The claim that externalities explain the territoriality of public goods begs the question, because externalities occur in discrete areas only because of the use of territoriality in the provision of public goods.

21. Sack, *Human Territoriality,* pp. 112–13.

22. E. H. Landon, *A Manual of Councils of the Holy Catholic Church* (Edinburgh: John Grant, 1909), 2:71.

23. The very existence of territoriality hinders movement and rearranges people and phenomena. For example, territories receiving funds from a national agency would attract the recipients of these funds to these territories and keep them there. Territoriality, then, has the effect of rearranging the distribution of phenomena as well as establishing different levels of access to goods and services.

24. For a review of community and its unattainability, see Glenn Tinder, *Community: Reflections on a Tragic Ideal* (Baton Rouge: Louisiana State University Press, 1980).

25. See Ferdinand Tonnies, *Community and Society* (East Lansing: Michigan State University Press, 1957); and Émile Durkheim, *The Division of Labor in Society* (Basingstoke, Hants.: Macmillan, 1933).

26. Ellen Semple, *Influences of Geographic Environment on the Basis of Ratzel's System of Anthropo-geography* (New York: Henry Holt, 1911), p. 115.

27. J. Nicholas Entrikin, *The Betweenness of Place: Toward a Geography of Modernity* (Baltimore: Johns Hopkins University Press, 1990), pp. 66–67.

28. MacIntyre, *After Virtue.*

29. M. I. Finley, *The World of Odysseus* (New York: Viking, 1954), p. 134, as quoted in ibid., p. 115.

30. MacIntyre, *After Virtue,* pp. 118–19.

31. Ibid., p. 127.

32. Ibid., p. 132.

33. Norman, *The Moral Philosophers,* p. 54.

34. Aristotle, *Nicomachean Ethics,* trans. W. D. Ross, bk. 4, chap. 1, p. 1120b, lines 3–4.

35. MacIntyre, *After Virtue,* p. 147.

36. David Miller, "The Ethical Significance of Nationality," *Ethics* 98 (1988): 647–62; quotation on p. 647.

37. Ibid., p. 649.

38. The phrasing is from Norman, *The Moral Philosophers,* p. 102, wherein he presents three phrasings of this imperative.

39. *The Autobiography of Benjamin Franklin* (New York: Collier Books, 1962). Franklin states that in his readings he found the catalogue of virtues numerous and inconsistent, and so for the sake of clarity he decided "to use rather more names with fewer ideas annex'd to each," p. 82.

40. See MacIntyre, *After Virtue,* p. 12.

41. Ibid., pp. 104, 105.

42. Tuan, *Morality and Imagination.*

43. Hans Jonas, *The Imperative of Responsibility: In Search of an Ethics for the Technological Age* (Chicago: University of Chicago Press, 1984), pp. 4–5.

44. Ibid., p. 1.

45. Ibid., p. 11. The same point can be rephrased: (1) Do not compromise earth's conditions for an indefinite continuation of humanity on earth; or (2) think globally, act locally.

46. Many effects can become physically encompassed by the use of territoriality. This means that activities become molded by territorial form, much like societies that can be territorially defined.

47. Susanna Hecht and Alexander Cockburn, *The Fate of the Forest: Developers, Destroyers, and Defenders of the Amazon* (London: Verso, 1989), is an excellent example of the way the geographical approach can provide a balanced analysis.

48. Colleen McDannell and Bernhard Lang, *Heaven: A History* (New Haven: Yale University Press, 1988).

49. Hannah Arendt, *The Human Condition: A Study of the Central Dilemmas Facing Modern Man* (Garden City, N.Y.: Doubleday, 1959), argues that modern consumption is much like metabolizing, hardly something that distinguishes humans from the rest of the animal kingdom.

50. A "growth industry" in magazines and popular publications has developed around increasing consumer awareness of the consequences of consumption. These industries do not attack consumption in general but, rather, disclose its effects on each of these realms. Certain products become more or less acceptable according to the rankings placed on them by these consumer advocate groups.

51. Marx and Engels offer only scattered statements about what was to succeed capitalism. Alfred Schmidt, *The Concept of Nature in Marx* (London: NLB, 1971), p. 144, presents an excellent summary:

> The marxist view of labour in the future "association of free men" can be formulated roughly as follows: men should not be oppressed in their labour . . . however, labour cannot altogether be abolished and replaced with what is now called leisure-time activity, in the course of which men senselessly waste their time and yet simultaneously remain bound to the rhythms and ideology of the world of labour. The free time of the future will not be merely a quantitative extension of what today is understood as free time; culture is not a fixed physical stock of things which will come into the possession of the "whole people" in more numerous and improved editions. Only when "immediate labour-time" ceases to stand "in abstract opposition to free time" can human qualities be universally unfolded and, in their turn, again work to further the growth of the forces of production.

As Marx puts it in the *Grundrisse* (599), as quoted by Schmidt:

> The saving of labour-time is the same as the increase of free time, i.e., time for the full development of the individual, which itself again works back as the greatest force of production upon the productive power of labour. . . .

> Free time—which is both leisure time and time for higher activity—has naturally changed its owner into another Subject, and as this other Subject he then enters directly into the process of production.

This results in an emancipation of the individual. Again from the *Grundrisse* (593):

> The free development of individualities is needed, and therefore not the reduction of necessary labour-time in order to replace it with surplus labour, but the reduction altogether of the necessary labour of society to a minimum which would suffice to create the means for the artistic, scientific, etc., education of the individuals in the time which had become free for them all.

Also of especial importance are the remarks of Karl Marx in volume 3 of *Das Kapital* (cited in Jonas, *The Imperative of Responsibility,* p. 193). "The realm of freedom begins indeed only at the point where labor determined by need and external expediency ceases. It lies therefore by the very nature of the matter beyond the sphere of material production as such." See also Marx's statement (ibid., p. 241) that "beyond [this residue of necessity] there begins the development of human power acting as its own end, the true realm of freedom." This type of work will lead to true human expression: "in place of the old bourgeois society, with its classes and class antagonism, we shall have an association in which the free development of each is the condition for the free development of all." (See Karl Marx, "Manifesto of the Communist Party," in Karl Marx and Friedrich Engels, *Selected Works* [Moscow: Foreign Languages Publishing House, 1955], 1:54.) See also Engels' statement at the end of *Socialism: Utopian and Scientific* (New York: International Publishers, 1935) about humanity's leap from the realm of necessity into the realm of freedom, which makes it appear that necessity is abolished absolutely, while Marx believes it is still there internalized.

52. Karl Marx and Friedrich Engels, *The German Ideology* (London: Lawrence & Wishart, 1965), p. 45, reads:

> for as soon as the distribution of labour comes into being, each man has a particular, exclusive, sphere of activity, which is forced upon him and from which he cannot escape. He is a hunter, a fisherman, a shepherd, or a critical critic, and must remain so if he does not want to lose his means of livelihood; while in communist society, where nobody has one exclusive sphere of activity, but each can become accomplished in any branch he wishes, society regulates the general production and thus makes it possible for me to do one thing today and another tomorrow, to hunt in the morning, fish in the afternoon, and rear cattle in the evening, criticize after dinner, just as I have a mind, without ever becoming hunter, fisherman, shepherd or critic.

53. Without the necessity of labor, our hobbies become our vocations; see Jonas, *The Imperative of Responsibility,* p. 195. Yi-Fu Tuan recognizes the importance of necessity for the good life: "Paradoxically, necessity gives ease to life. Necessity, not only as a fact recognized by the mind but as a weight fully felt by

body and soul, sustains the objective character of the world." *Segmented Worlds and Self: Group Life and Individual Consciousness* (Minneapolis: University of Minnesota Press, 1982), p. 197.

54. Mihaly Csikszentmihalyi and Isabella Csikszentmihalyi, *Flow: The Psychology of Optimal Experience* (New York: Harper and Row, 1980).

Author Index

Numbers in italics indicate page of citation in the notes.

Subject Index

About the Author

Robert David Sack was born in Brooklyn, New York, and was educated at the University of Pennsylvania (B.A., political science) and the University of Minnesota (M.A. and Ph.D., geography). Since 1970 he has taught at the University of Wisconsin—Madison, where he is professor of geography and professor of integrated liberal studies. He has received numerous grants, including a John Simon Guggenheim Memorial Fellowship and a Fulbright Research Fellowship to England. He is the author of *Human Territoriality: Its Theory and History* and *Conceptions of Space in Social Thought.*